SNOWBALL ORANGES

ONE MALLORCAN WINTER

Peter Kerr

SUMMERSDALE

Summersdale Publishers Ltd
46 West Street
Chichester
West Sussex
PO19 1RP
UK

www.summersdale.com

Printed and bound in Great Britain.

ISBN 1 84024 112 8

Front cover photograph: The Travel Library – David Toase
Cover design: Blue Lemon

CONTENTS

For Majorca is one of the most
beautiful places on Earth . . .
Like a green Helvetia
under a Calabrian sky,
with the solemnity and silence
of the Orient.

George Sand, 'A Winter in Majorca' – 1838–1839 (Luis Ripoll, 1988)

– ONE –

'SE VENDE'

'Look! The weather has come from Scotland to welcome you to Mallorca,' beamed Tomàs Ferrer, blinking in delight at the rare sight of snow falling from Mediterranean skies onto the trees of ripening oranges.

'*Sí, sí,* I see it,' was all I could say, my courteous attempts to share our new neighbour's enthusiasm failing with each fluffy flake which landed on my upturned face. Surely we hadn't migrated fifteen hundred miles southwards only to be followed by the same freezing weather which we had just left behind?

Señor Ferrer wandered through the orchard, kicking up the snow and chuckling with childlike glee. I could hardly believe my eyes. A cold mantle of white was rapidly transforming our sunny paradise into a bizarre winterscape of citrus Christmas trees, cotton wool palms and snowball oranges.

* * * * * * * * *

It had all looked so different only a few hours earlier when Ellie and I pushed open the shutters to reveal the trim rows of fruit trees lining the valley floor, already bathed in warm December sunshine. It was our first morning on our newly-acquired Mallorcan farm, and we were content just to stand silently by the open window, allowing our senses to be intoxicated by the magic brews of sunlight and shadow, by drifting scents of wild herbs and the soothing sounds of tinkling sheep bells which fill those secret wedges of fertile land nestling so peacefully between the deep folds of pine-clad mountains in south-west Mallorca.

This was truly an enchanted island, and we were blessed to be looking out over our own little piece of it – a few acres of sun-kissed heaven.

A shrill *'SEÑOR P-E-T-E-R . . . SEÑOR P-E-T-E-R . . . H-O-L-A-A!'* sliced through the tranquillity like a blunt knife.

I opened the front door to behold the diminutive silhouette of a woman surrounded by a halo of golden light. A host of smaller silhouettes with ragged, spiky outlines milled around her sandalled feet, and there was something so strangely saint-like about this vision, blending almost hypnotically with the feeling of abiding calm which pervaded the valley, that I was all but overcome by a compelling impulse to drop to my knees in obeisant homage.

I was brought back to my senses abruptly and painfully, however, by one of the little silhouettes levitating swiftly and embedding its claws firmly in my right leg. We were being paid our first visit by Tomàs Ferrer's wife, Francisca, and her group of faithful disciples – eight moth-eaten, half-wild cats and two skulking mongrel dogs.

The surprise cat attack on my leg turned out to be merely a diversionary tactic, allowing the rest of Señora Ferrer's hairy *bandidos* to dive past me and disappear in all directions throughout the house.

Señora Ferrer giggled a coy *'Bon dia.'*

I managed a pained *'Buenos días'* in reply, at the same time struggling to remove her fiendish familiar from my thigh, claw by determined claw.

'Qué gato tan malo!' Señora Ferrer chided gently, whipping the hissing rat-bag off my leg with a deft flip of her sandal.

The cat, mission accomplished, shot off, tail high, in search of the rest of its gang, who were now, I feared, occupying the entire house.

This was our first meeting with our new neighbour since concluding the purchase of the farm from her some five months earlier.

Although not accepted practice in Britain, both the Ferrers and ourselves had been represented during the lengthy and tortuous sale/purchase procedures by the same lawyer, a tall, laconic mainland Spaniard who spoke excellent English.

'I must warn you about doing business with Mallorcans,' he'd said on my first meeting with him, revealing to me for the first time the mutual animosity that exists between certain mainland Spaniards and their island-born countrymen. 'You will find the typical Mallorcan a wily negotiator, not willing to give a centimetre, or, more importantly, a peseta. He applies the same attitude to significant property transactions today as his forefathers did to haggling over the price of an old hen at market.'

The lawyer had allowed himself a wry smile when I then observed that the Mallorcans, therefore, clearly had a lot in common with the Scots.

'A Mallorcan and a Scotsman doing business,' he'd muttered through puckered lips, arching his eyebrows as he commenced his scrutiny of the deeds to the Ferrers' farm. 'Hmm, *vamos a ver*. It will be interesting.'

And true enough, the Ferrers *had* gone on to display a typical country canniness in their dealings with Ellie and me, but no more than we were used to at home, and not enough to prevent them finally accepting a price for the farm which was well below that originally asked. But any smugness which we might have felt about 'out-wilying' a Mallorcan was well and truly tempered when we ultimately discovered that the Ferrers had actually conducted negotiations in the accepted Mallorcan way, by asking a considerably higher price than they'd really be prepared to accept . . . if pushed. We awarded ourselves at best a draw, a gesture which all too soon we were obliged to concede may have been just a tad generous.

But there had been little time to entertain such misgivings during the ensuing few months, which were spent in a flurry of activity, arranging and concluding the sale of our own farm in Scotland, organising immigration permissions from the Spanish Consulate, having our 'good characters' certified by a notary for the Spanish military who, 'for reasons of national security', had to vet each foreigner wishing to buy rural land in the Balearic Islands, arguing about which non-essential bits and pieces should or should not be shipped to Mallorca, and finally saying our goodbyes to family and friends.

Our two sons, Sandy (eighteen) and Charlie (twelve), would join us in a few weeks, when we hoped to be fully settled in. And although we couldn't help but have nagging doubts about how the boys might eventually react to such a total change in environment and lifestyle, it was too late now. We had crossed the Rubicon, for better or for worse . . .

Like so many others, we had often passed a bleak winter's evening indulging in dreams of someday swapping the rigours of growing crops and raising cattle in the harsh Scottish climate for the unknown but decidedly more attractive rigours or running some sort of small farm in Spain . . . someday, when our two sons were older and settled into their eventual careers . . . someday, when we could really afford to take the risk . . . someday, when we had found time to learn Spanish properly . . . someday.

No doubt, that someday would never have come had we not literally bumped into the farm of 'Ca's Mayoral' during a summer holiday in Mallorca.

We were driving north from the market town of Andratx, intending to escape the blistering heat of the coast by taking the road up through the mountains to the village of Capdella. But instead, Ellie's unique navigational skills had led us into a hidden valley not even marked on the map, down a narrow lane meandering between sun-baked drystone walls which enclosed little fields of almond trees, their spidery branches and sparse foliage providing a porous canopy of shelter from the fierce July sun for the small groups of skinny, lop-eared sheep lying panting at the foot of the ancient, gnarled trunks.

The lane was only just wide enough to take the car, and by the time we had realised our mistake, I was faced with the choice of either driving on in the hope of finding a turning place, or trying to reverse all the tortuous way back to the main road. I decided to drive on . . . and on . . . and on . . .

'How the hell can you *still* not read a simple map?' I eventually blurted out. 'I mean, there *is* only one real road in these parts, and yet you've managed to lose it and get us stuffed down this God-forsaken donkey track on the road to nowhere. I've said it before and I'll say it again — I really would be better off reading the bloody road map myself *and* driving at the same time.' The heat was getting to me, and besides, I was warming to my theme. 'You go through life with your eyes half-shut, Ellie, that's your trouble. After all, there's nothing very difficult about reading a map. It's only a matter of keeping your eyes open and being observant, that's all!'

Ellie replied with a practised, smiling silence.

I piloted on, muttering at the sun while the passing chirrup of basking grasshoppers rose and fell on the oven-hot air from parched clumps of withered weeds rasping against the sides of the car with brittle and lifeless leaves.

'There's a gateway in that big wall — just up ahead on the left,' Ellie indicated after a few minutes. 'You can turn there.'

'I know, I know!' I snapped, screeching to a halt several yards past the turning place.

Ellie allowed herself a knowing smirk. 'Look at that sign,' she said as I reversed towards the opening.

'What sign?'

'The one on the gate that you just backed into, Mr Observant.'

I remained prudently quiet for a moment or two, looking over my shoulder at the large *'SE VENDE'* sign. 'U-huh —

For Sale,' I shrugged. 'So what? It'll probably be one of those tiny postage stamp fields with a few stunted trees and a tumble-down stone shack in the corner like all the other ones we've passed on the way down this damned track. Let's get the hell out of here before we melt.'

'So you didn't notice the house?'

'Come again?'

'Did you not *observe* the house? That's what I'm saying.'

'What house?'

'That big, white farmhouse sitting among all the fruit trees. You couldn't miss it. It was right there, just a hundred yards in front of us as you drove round that last bend. *This* is its gate. The *house* is in there. See?'

I cleared my throat. 'Oh *that* house. Yes, very pretty. Anyway, let's get back on the road to Capdella.'

'No, wait a minute,' urged Ellie. 'There's a woman in there watering the garden. Let's go and ask her to show us round. It *is* for sale, so why not have a look?'

I started to suggest that there would be no point and that it would only be wasting the woman's time, but Ellie was already through the gate and engaged in an animated exchange of one-word and two-word sentences.

As I approached them, it became obvious that the watering lady could speak no English at all, and as Ellie's Spanish was only just approaching supermarket standard, I fancied that there might be an opening here for my 'superior' linguistic skills.

'Ah, ehm . . . *buenos tardes, señora*. Ehm . . . *como está usted?*' I bumbled fluently, thrusting out my hand in greeting.

'*Hola,*' the woman smiled nervously. '*Soy Francisca Ferrer.*' Her right hand moved hesitantly forward to meet mine, and I felt a wet sensation spreading down the front of my trousers. She was still holding the hose-pipe.

Ellie rolled her eyes heavenwards. 'Just leave the talking to me, dear,' she sighed.

Señora Ferrer offered profuse apologies, but could not suppress a fit of girlish sniggers. She was obviously a woman with a sense of humour, but one I was unable fully to appreciate at that particular moment. My drenched trousers clung uncomfortably to my legs, dredging up sorry memories of a once-familiar sensation that I hadn't experienced since my early days at infant school.

I waddled stiffly behind Ellie and the Señora as they re-immersed themselves instinctively in their mutual language of shouted monosyllables and flamboyant hand gestures. My wife was soon able to inform me that this was the *finca* – or farm – of Ca's Mayoral, the birthplace of Francisca Ferrer, and her family's ancestral home.

Señora Ferrer was a fairly small woman in her mid-fifties, I guessed, and although she was native to this valley of little farms, she had an appearance untypical of most women who spent their lives working in the fields. She had smooth, delicate hands, immaculately coiffeured hair, and an almost regal bearing which seemed to add inches to her modest height as she strolled elegantly between exotic shrubs and flowers towards the house.

It ultimately transpired that she was the wife of Tomàs Ferrer, an upwardly-mobile man of his time who, in his youth, had left his own peasant family's land on the other side of the island to embark upon a career in local government in Mallorca's capital city, Palma. He progressed, and was subsequently posted some twenty miles away to the country town of Andratx, where he met and later married the young Francisca, only daughter of one of the district's most respected fruit farmers. Tomàs soon became an important official in the area and, in the fullness of time, he was recalled to Palma, where his proven administrative skills finally elevated him to the position of *El Director* of one of the island's most vital *ministerios*.

The Ferrer's success story had become part of the folklore of Andratx, where Tomàs and his wife were still held in great esteem, particularly by the people of Francisca's valley – a status in which the Señora clearly revelled.

During the week, the Ferrers lived in a smart apartment on one of Palma's most fashionable tree-lined *avenidas*, returning to Ca's Mayoral every Friday evening to spend the weekend with Francisca's ageing parents. Despite their fairly high position on Palma's social ladder, the peasant blood still ran strong in the Ferrer's veins, and they delighted in steeping themselves in the almost extinct rural lifestyle of their childhood for those two precious days each week.

Tomàs would help his father-in-law with tractor work and some of the heavier farm chores that were becoming too much for the old man. Meanwhile, Francisca preferred to pass the time of day tending the flower beds near the house, where she could gossip in a suitably genteel manner with the old peasant women who walked up from the village to buy fruit at the farm gate and to hear the latest Palma scandal first-hand from *La Condesa*. Her mother would divide her time between the kitchen and the fields in the age-old way of Mallorcan countrywomen, tirelessly and lovingly tending the needs of family and farm, yet always finding the time for a cheery chat with passing friends and neighbours.

Then the old lady died, and not only had the hub of the little family been taken away, but Francisca's broken-hearted father began to lose interest in his once-cherished trees and land. Without his life's partner, nothing could ever be the same again.

The old man's health was already poor, and Francisca worried about leaving him alone all week in that big farmhouse with no one to look after him. There was nothing else for it – he would come to live with them in Palma. The farm would be sold, except for one field by the lane where there was an old mill which they could convert into a *casita fin de semana*, a weekend retreat to which they would return to enjoy the relaxation of growing fruit and vegetables, and where the old fellow could wander beneath the fruit trees and quietly recall a lifetime of memories in his beloved valley. The *'Se Vende'* sign was reluctantly posted.

Ellie and Señora Ferrer entered the house, still conversing in their own improvised *Esperanto*, but I lingered outside for a while – partly to dry off my sodden trousers, but mainly just to savour the wonderful ambience of the place. An irresistible feeling of timeless peace was everywhere, a still silence broken only by the drowsy chirping of sparrows in the tall pines by the old wrought-iron gate.

I was standing in the yard behind the house; a gently sloping gravelled area, entirely hidden from the outside world by a massive high wall of ancient beige sandstone. Billows of purple bougainvillaea cascaded over its craggy face, the long, leafy tendrils weaving intricate patterns with their shadows against the cracked blocks of sun-beaten stone. The other side of the yard was bound by a pretty copse of flowering hibiscus, oleander, mimosa and climbing roses, living in a wonderfully abandoned harmony with a few olive, almond and carob trees which looked as if they had occupied the site for many centuries longer than their more colourful bedmates. It seemed that Señora Ferrer's hand (and her hose-pipe) had helped Mother Nature create a visual masterpiece.

The yard narrowed towards the far corner of the house, and I had to stoop beneath drooping palm fronds and a tangle of wisteria to reach the west-facing terrace, shaded from the afternoon sun by a pergola of wooden beams draped in vines. Bunches of tiny young grapes were already hanging down through the translucent canopy which cast shimmering shapes of refreshing verdant light over the stonework of an old well standing foursquare below. But unlike the 'wishing' variety so popular in manicured suburban gardens, this old well had no quaint thatched roof, just a sturdy arch of weathered iron with a rusty pulley, a dilapidated wooden bucket dangling lazily on the end of a length of age-bleached rope.

Opposite the well, two rickety wooden chairs leaned eccentrically against the house wall, surrounded by a glorious scatter of hessian sacks, assorted fruit trays, an earthenware water jug and a clutch of empty wine bottles. A fat hen clucked contentedly in a battered wicker basket under one of the chairs, confirming my suspicion that this terrace was a cool oasis where afternoon work might be undertaken – if absolutely necessary – but only the type of work which could best be tackled sitting down. The notion appealed to me.

Emerging into the full sunshine from the dappled light of the terrace, I had to shield my eyes against the glare from the house's whitewashed walls. The heat was intense – a bit too ferocious for my unaccustomed northern hide, but just right, it appeared, for a small colony of cicadas screaming happily in the swaying branches of two eucalyptus trees standing guard over the front of the building. I made for the long, open-fronted *porche* that sheltered the main entrance, and as I reached the welcome refuge of its shade, I could hear Ellie and Francisca Ferrer nattering away loudly behind the arched double doors.

'Electricity or gas . . . the cooker, the oven . . . you know . . . it's gas, yes?'

'Ah, sí – butano – butano gas en botella.'

As their conversation had arrived at the basic female topic of cooking facilities, I suspected that a serious property-purchase negotiation was either already in progress or dangerously imminent. Rapid preventive action was called for. I knocked urgently on the parched oak door, and it creaked open to reveal two smiling faces peering from the shadowy interior.

'Aha, señor. Por favor,' welcomed Señora Ferrer, ushering me keenly inside.

'Peter, you must come in and look round this gorgeous old house,' enthused Ellie, grabbing me by the elbow and brushing aside any attempted protestations as she whisked me on a lightning tour of every room.

I staggered back into the sunlight a few minutes later with no detailed recollection of what I had just seen – only blurred images of airy white rooms with shuttered windows and beamed ceilings, a stone staircase here, a big open fireplace there, an overpowering smell of cats, and the ever-present growling of unseen but patently unfriendly dogs.

'Oo-ooh! El sol, siempre el sol. Es terr-ee-ee-blay, horr-ee-ee-blay,' wailed Señora Ferrer, throwing her forearm over her brow in exaggerated protest at the heat which hit us as we stepped out from the *porche* and headed for the fields. Ellie and I were used to moaning perpetually about the biting cold and driving rain of endless Scottish winters, so it was odd to hear someone actually describing the sun as 'terrible, horrible'. Some people are never pleased, I thought to myself, rubbing the stinging streams of sweat from my eyes.

For the next half-hour, we followed our chattering guide through lines of unfamiliar fruit trees, only half listening to her rambling descriptions of pomegranate and quince varieties, spray and fertiliser requirements, harvesting techniques and irrigation timings – all of which would still have seemed like double Dutch to us, even if she'd been speaking in English. Barley and bullocks we understood, lemons and loquats we didn't.

But the valley was already casting its spell, and my resistance was on the wane. As we moved from one little stonewalled field to another, our eyes were constantly drawn to the ever-changing aspects of the mountains which rose majestically on either side. Their steep slopes were covered in a rich green carpet of pine and evergreen oak, the trees clinging precariously to the narrowest of ledges all the way up to the *sierras altas*, the saw-toothed ridges high above where earth meets sky. Rocky outcrops shimmered pink in the warm glow of the sun – sheer cliffs and dark gorges completing a picture of wild natural beauty both contrasting with and complementing the patchwork of tiny fields and orderly plantations lying far below. We were enveloped in an almost tangible tranquillity that seemed to generate a feeling of belonging in that near-perfect blend of the works of nature and man.

We gazed back over the fields to the farmhouse. Its white walls, faded wooden shutters and terracotta tiled roof peeped sleepily over the deep green domes of orange trees, while the mountains looked benignly on – secure, solid and serene. Without exchanging a word, we both knew that this was going to be our new home.

On the way back to the farmstead, what seemed like a whole herd of mangy cats materialised one by one as if from nowhere to follow Señora Ferrer, and as we neared the house, two scruffy dogs bounded from the front door, ignoring Ellie and me totally in a leaping, yelping, tail-wagging welcome to Francisca, their returning leader.

We thanked Señora Ferrer for her time, informed her of our interest in buying the property, and promised to return next day with a bi-lingual lawyer. This piece of news sent the Señora into an excited flutter of two-cheek kisses and (thankfully) hoseless handshakes. Backing towards the car under a barrage of *hasta mañanas*, we finally managed to say goodbye.

The cats couldn't have cared less. They were too busy jostling for rubbing positions around their mistress's ankles. But the two dogs were clearly delighted to see us leave. Until then, they had only appeared threatening from a strategically safe distance, but that wary hostility turned to matador-style confidence as soon as we got into the car and closed the doors. The larger mutt strutted forward, gracefully raised a leg and, with a disdainful toss of his head, peed a bucketful over our offside rear wheel. Meanwhile, his little pal posed boldly in front of the car and haughtily back-kicked sprays of gravel over the bonnet, as if covering a freshly laid turd.

Olé, and *Adiós, amigós!*

The battle lines had been drawn by the dogs back on that fateful July afternoon, but now that we had actually moved in to the house, the cats appeared to have assumed responsibility for letting us know that we were not welcome in *their* home.

Francisca had brought us a housewarming gift of some of her home-baked almond cake, made from nuts from the trees at the side of the house, she assured us. And to add to the treat, she pulled from her basket a tub of her very own almond ice cream, made to a secret family recipe which had been handed down from generation to generation. This ice cream was *muy especial* and could not be obtained anywhere else, she stressed. We would never have tasted anything like *this* before. How true.

We sat down at the kitchen table and watched Señora Ferrer apportioning fair shares of her goodies proudly and deliberately, as if weighing out the most expensive beluga caviar. We prepared our tastebuds for amazement.

My mouth had scarcely closed on the first bite of almond cake when the Señora's face appeared right in front of mine, our noses almost touching, her eyes wide with anticipation. *'Muy buena, la tarta, no?'*

'Very good . . . yes,' I spluttered, pieces of flaked almond shooting out of my mouth and fixing themselves firmly on the lenses of Francisca's spectacles.

Ah sí, it was a little dry, *la tarta,* if eaten without the ice cream, she explained. It was necessary to take equal amounts of cake and ice cream on the spoon at the same time. She gave me a slow-motion demonstration. Now *I* must try.

I felt like an overgrown baby being given lesson *numero uno* in basic spoonmanship. Without even glancing at her, I knew that Ellie was trying desperately hard not to laugh, and that didn't help to ease my embarrassment one little bit.

The combined spoonful technique certainly made the dust-dry cake easier to masticate, but despite Francisca's big build-up, I thought her secret almond ice cream just tasted like watery milk laced with powdered nuts. But, in the interest of politeness, I couldn't let her see me gagging. I had to persevere. Maybe this was an acquired taste, I thought. Indeed, how would a Mallorcan pharynx react the first time it was confronted by a dollop of good old Scottish porridge?

Señora Ferrer's expectant face waited for a sign of approbation. Realising that there was no escape, I took a deep breath and swallowed hard. 'Mm-mm-mm-mm,' I crooned, nodding my head and making a brave attempt at lip-smacking. I was learning to lie in Spanish without even speaking. My well-intentioned dishonesty was rewarded with another generous helping of the dual confectionery masterpiece. I had made Francisca's day.

Just then, I became aware of a frenzied scratching noise coming from under the table and I felt a series of sharp needles scraping down my shin. It took a few seconds for the stark truth to sink in – but yes, one of her cats was actually sharpening its claws on my leg!

'Ouch!' I yelled, pointing frantically at my disintegrating trouser leg. 'The cat, your cat . . . it's . . . it's . . .'

'*Ah sí, es normal,*' smiled Señora Ferrer calmly, adding that it was really the ice cream that the little dear was after. It had a weakness for her ice cream, that one. Oo-ooh, and it was purring too. It liked me. It *really* liked me. She clapped her hands to her cheeks in delight, inclining her head to one side in dewy-eyed emotion at the sight of this violent love scene.

My mind was racing. My shin was bleeding. Then my besieged brain came up with one beautifully simple solution to my two pressing problems. I snatched my brimming plate off the table and transferred it rapidly to the floor. It did the trick. In an instant, my sadistic 'admirer' was off my leg and was lapping furiously at the ice cream. As if by magic, the cat's fellow gang members appeared from their hiding places, and suddenly the pudding plate was buried under a heaving circle of waving tails and winking ani.

Señora Ferrer was visibly overcome.

At first, I was afraid that I'd offended her. But not a bit of it.

Madre mía, what a kind man I was to give my ice cream to her precious little darlings, she simpered. She and Tomàs had not been gifted with children, so the cats and dogs were really their *niños*. I would understand, she was sure, because I was clearly an animal lover. I had a *magnífico* rapport with cats. I was a real cat person. *Sí, claro!*

Fair enough, I did like cats all right, but at that particular moment, I felt more like a boot-up-the-cat's-arse person. I had had one pair of perfectly good trousers and two reasonably good legs mutilated in the space of twenty minutes, and I knew deep down that what these cats really wanted was us out of the house – *their* house. Nevertheless,

I couldn't bring myself to tell Francisca that, and she probably wouldn't have listened anyway.

She continued to lavish praise on my 'obvious love for *animales*'. Her patron saint from Assisi would have been proud of me, she said, then paused and crossed herself pensively.

I sensed then that I was being softened up for a patsy punch – being sweet-talked into something that I'd live to regret.

Francisca's eyelashes fluttered behind her Catwoman glasses. '*Señor Peter . . . oh eh, y Señora,*' she said, flashing a diplomatic smile in Ellie's direction, then concentrating her aim on the easy target. Would I be so kind as to feed the cats and dogs on weekdays when she and Tomàs were away living in Palma? The cats being cats were quite *independiente*, as I would know, and they would probably be happy to sleep in one of the outhouses. The dogs – they were mother and son, by the way – Robin and Marian, after Robin Hood in *Inglaterra* – what a lovely coincidence that they had English names – well, the dogs would spend most days hunting in the mountains, but at night . . . Robin was very sensitive, *un poquito tímido*, and his little mother was getting old and *enferma* with arthritis, the occasional false pregnancy, a few cysts, that sort of thing . . . perhaps we would allow them to sleep in the kitchen as usual? She would provide us with their food and they could live with her over at the *casita* at weekends. Her little ones would give us *absolutamente no problemas*.

I didn't have the heart or, more accurately, the vocabulary to refuse. Señora Ferrer could not have been more delighted had she just won first prize in the *lotería*. She gave us both a big hug, and there were tears in her eyes. Ellie gave me a surreptitious kick on my ripped shin, and there were tears in my eyes too.

Promising to bring us some more almond cake and ice cream very soon, a contented Francisca assembled her troops and shooed them out of the front door, the merest glimpse of her raised sandal dissuading any maverick that dared to even think about sneaking back inside. The motley procession moved off, and the head lady shouted back that she would return in the afternoon to show Ellie how everything about the house worked. Tomàs would be coming too; there was a lot to explain to me about trees and things.

Robin and Marian were last to leave, creeping reluctantly from their treasured kitchen and glancing furtively at Ellie and me as they sidled past, tails firmly between their legs. At least they hadn't growled this time, though. Perhaps they were beginning to accept us already?

With this benevolent thought in mind, we set about unpacking the few suitcases that had arrived on the plane with us the previous night. The rest of our things wouldn't be delivered by the shipper for at least a couple of weeks. But not to worry. As is the custom in Mallorca, the house had been left 'furnished' by the previous owners. In truth, this meant that we had inherited a lumpy old bed with a grotesque headboard which was fighting for survival against an army of voracious woodworm; a tiny, threadbare three-piece suite which must have been designed for comfort by a church pew-builder; and a wobbly kitchen table and chairs that had definitely lost their battle against the woodworm years ago. A basic survival kit of some chipped crockery and Uri Geller-type cutlery completed the 'furnishings'.

None of this temporary discomfort mattered, however. We had made the big move, and we were thrilled to be the proud *dueños* of our own Mallorcan farm – a dream for many, but a dream which we had actually turned into reality. Flushed with self-satisfaction, we started a leisurely exploration of the house. I began to make a mental note of all the things which would need to be done in the way of repairs and improvements, but although the old place looked a bit neglected, there didn't seem to be any major defects at all. Certainly, it could have done with a lick of paint here and there, there were a few cracks in the plasterwork to make good, some broken floor tiles to be replaced and a couple of window shutters to be fixed, but nothing too serious, I decided. The main thing was that the house had a really friendly feeling about it, and I was beginning to like it more and more.

'You know,' I said, descending the little staircase into the kitchen, 'you were right about this place from the very first moment you saw it, Ellie.' I gave the old wooden bannister rail a pat. 'Yes, there's a good atmosphere in this house – good and welcoming. A lot of nice things must have happened in here. You can tell. It's the same with all houses – the vibes are either right or wrong, and I'm getting good vibrations here.'

'Glad to hear that you've come round to my way of thinking. Mind you, it would have been a bit too late if you hadn't. Come on, let's see if we can rustle up a cup of coffee on that old cooker.'

'Good idea. But do me a favour – don't give me any more of her ladyship's almond cake. The best thing for that stuff would be to feed it to the dogs.'

'Mm-mm, and talking of the dogs, I think they've left a little message for us,' Ellie said, screwing up her nose and pointing under the table.

And sure enough – there were Robin and Marian's calling cards, placed side-by-side on the floor . . . one lumpy, one liquid. All of a sudden, the atmosphere wasn't quite so welcoming. We got the dogs' message loud and clear, though. Like the cats, they wanted us OUT, and although we didn't know it, the friendly old house had a few surprises in store for us too.

* * * * * * * * * *

The freak snowfall put an end to Tomàs Ferrer's meticulous lecture on the intricacies of Mallorcan fruit farming. I'd had enough for one session anyway. My newly emigrated head was already throbbing from a concentrated bombardment of horticultural lessons in Spanish, my lacerated legs hurt, I had overdosed on almond ice cream, and the dogs and cats hated me. I was knackered.

Señor Ferrer shouted *adiós* and hurried off through the orange grove towards his *casita* – perhaps to indulge in some winter sports with Francisca before the inevitable quick thaw set in. I didn't know, and I didn't care.

I flopped onto an uncomfortable chair in the kitchen that now reeked of agricultural disinfectant. Ellie said something about Señora Ferrer insisting on spraying the room after Robin's and Marian's 'unfortunate tummy upset'.

'I never really wanted to come to bloody Mallorca in the first place,' I muttered to myself, descending into a wallow of self-pity. 'This place stinks like a public lavatory,' I declared to Ellie's rear end while she rummaged noisily in the nether regions of the cooker.

She stood up, sighed deeply, and turned round slowly. She was clearly in no mood to pander to any of my ill-natured mumblings. 'The gas bottle in the cooker is empty, it's Sunday night, and the *butano* truck won't be around again until Wednesday. We'll have to eat out. Let's go.'

I didn't argue. In fact, nothing was said during the two-mile drive into Andratx. Most of the snow had already melted, but the roads were wet, and slush lingered in the gutters of the town's deserted streets. The locals were snugly indoors, I reasoned – probably laughing in large family groups gathered round dinner tables groaning under huge platefuls of hearty country fare. It was dark, and the dim streetlights were reflected in puddles that trembled in the increasingly cold north wind. The dreaded *Tramuntana* was coming.

We cruised slowly through the old quarter of town, past a couple of charmless, fluorescent-lit bars with steamed-up windows, but no welcoming *restaurante* signs were to be seen. It had been a long time since our last hot meal on the flight over; now hunger and fatigue were growing into despair.

'Oh for a big bag of fish and chips,' I groaned. But this was rural Spain, and such homely delicacies were simply unavailable – except in some of the infamous holiday resorts along the coast maybe? No, we weren't *that* desperate.

A black cat ran across the street in front of us and darted into a narrow alleyway. Ellie peered after it, then nudged my arm. 'Follow that cat. There's a restaurant up there, I'm sure.'

We climbed the steep, cobbled lane to where the cat was sitting miaowing under a creaking hand-painted sign – 'Ca'na Pau – Restaurante Mallorquín'. Bingo! We had found a source of sustenance at last, and we had been guided to it by a cat, of all things. The door opened, the black cat scampered in and six other cats ran out. We couldn't recognise any of them as being members of Señora Ferrer's band of feline thugs, so we ventured inside.

'I am Pere Pau, *el patrón*,' rasped a voice as rough as sandpaper. *'Benvinguts a mi casa.'* Our host stood gaunt and inscrutable inside the open door – a bony Jack Sprat of a man with the sceptical leer and calculating eyes of a Dickensian cutpurse. He beckoned us with a wide sweep of his arm into the smallest eating place I had ever seen. The glazed door clattered shut behind us, and Señor Pau grabbed my hand in a finger-crushing shake. Ellie cringed behind me. He was dressed in full chef's rig-out, topped by the tallest of tall white hats which almost dusted the exposed beams of the low ceiling, a bent cigarette dangling from his protruding lower lip, and a tortoiseshell cat tucked comfortably under his left arm. He was already blocking the only visible escape route, so we were obliged to follow his direction to one of only five tables that were crammed into the tiny room. Becoming ominously aware that we were Pere Pau's only clients, we had to fight the feeling of rising panic that must grip every insect when it suddenly realises that it has flown smack into a spider's web.

The tables were draped in blue-and-white checked oilcloth, adorned only with a centrepiece ashtray emblazoned with the San Miguel beer logo. Harsh strip lighting illuminated the yellowed walls which were covered in an untidy collage of faded photos of past Andratx football teams, dog-eared posters announcing local sporting fixtures, and el cine Argentina's programmes of coming film attractions – all of them years out of date. We seemed to have stumbled into some kind of cranky time warp.

On either side of the door stood two tall fridge-freezers, their scratched white paint festooned with stickers advertising everything from fresh milk to drain-cleaner. A wooden stair, roped off with a Privado sign (yet going nowhere), rose from a dark corner of the room, and opposite was the open-plan kitchen – a cramped alcove half-hidden behind a high counter which was cluttered with heaps of assorted plates and a jumble of cutlery. Somewhere, hidden in the kitchen recess, a transistor radio honked out the latest Spanish pop-rock music . . . 'BABY, BABY, SOY UN ANIMAL, UN ANIMAL PARA SU AMOR – PARA SU AMOR, UN ANIMAL CRAZEE-EE PARA SU AMOR . . . YEAH!' An unseen cat wailed in unison.

We were seriously tempted then to make a beeline for the door as soon as Pere Pau's back was turned. But what the hell. We were ravenous, and what was there to choose between dying of starvation and dying of food poisoning? We asked to see the menu.

'Menu? La Carta? No, no, no!' stormed chef Pau. No need for such frills! He made two completely different dishes each day – but no choice. You took what you got. And no sweet either, he emphasised. After two platos of his food, no one had room for postre, unless they had worms. Of course, the Germans usually demanded a sweet, he confided

30

as an afterthought, but even they had to make do with an ice cream from the freezer.

And what were today's *platos*, I ventured to ask?

He dragged deeply on his cigarette, which was now so small that he could barely hold its soggy remains between the tips of his forefinger and thumb. For *el primero*, he declared, there was *Sopa de Pescado* – one of his *famoso* specialities, and well known as the best fish soup in the whole of Spain – and we should consider ourselves fortunate that it was possible to sample it at all tonight, because his good fisherman *amigos* down at the Port of Andratx had only managed to haul in a very meagre catch before today's storm broke.

We visualised being confronted with a smelly concoction boiled up from the previous few days' fishy leftovers – maybe even stuff which his cats had rejected.

'Then for the main course?' I immediately wished that I hadn't bothered to ask, but Pere had us where he wanted us and he was not to be denied now.

'Ah, *el segundo* – un plato típico de Mallorca . . . El Lomo con Col! *Fantástico! Estupendo! Mm-mm-wah!*' He threw his head back, intending to kiss his fingertips in true chef style, but succeeding only in grinding his glowing cigarette end into his melodramatically pursed lips. '*MIERDA!*' he cursed, dropping the burning butt which fell like a well-aimed incendiary bomb right into the turned-up cuff of his left trouser leg.

Several chairs were overturned in the wake of his mad dash towards the kitchen, and although we couldn't see what ensued behind the high counter, the sounds of Pere's demented flamenco dance and the splash of copious quantities of water being hurled around conjured up vivid images of utter bedlam. By way of confirmation, the panic-stricken tortoiseshell cat – screeching like a banshee – bolted for cover under Ellie's seat. Could we have wandered, perhaps, into the canteen of the local lunatic asylum?

Outwardly unperturbed, Pere made his reappearance. He trimmed the angle of his chef's hat, dabbed his blistered lip with his apron, shook the water from his left shoe, and squelched confidently towards us, head held high. His dignity was being maintained . . . at all costs.

Would we care for a drink while the food was being prepared, he enquired? (It was now quite clear that Señor Pau was maître d', chef and headwaiter all in one – and possibly dishwasher too.) We were informed solemnly that there *was* a choice of wine. We could have either red or white. I opted for the red, and Ellie asked if she could have a coffee, *un café solo*, to warm her up.

An expression of shock spread over Pere's face. *'Un café?' No es posible!'* he retorted, visibly hurt. If anyone wanted a coffee, they should go to a bar. This was a restaurant, *his* restaurant, and the purpose of the establishment was to allow his clients to enjoy *his* food, his fine food, made only from the best local produce – all *totalmente natural* and with no nasty preservatives. No, it would *not* be possible to have a coffee. A chef does not make coffee. He *creates* food . . . *good* food. *Hombre!*

Ellie ordered a glass of water instead.

A large earthenware jug of red wine and a bottle of mineral water were duly delivered to our table with a generous bowl of olives and a basket brimful of crusty fresh bread. The transistor radio was mercifully switched off, his cats were

32

unceremoniously ejected from the premises, and our chef retired to his kitchen to create.

'I cannot stand distractions when I am working,' he announced, disappearing behind the kitchen counter in a cloud of smoke from a newly-lit cigarette.

We wolfed into the bread and olives while the sound of Pere singing and the surprisingly tantalising aroma of his cooking wafted over us. The wine was strong, earthy and good – from Binissalem in the middle of the island, our host informed us, poking his head out of the kitchen to check on our welfare, or, as Ellie remarked dryly, to make sure we were still there. Whatever, the wine was from the *bodega* of another of Pere Pau's *amigos* 'who always saw him right' – and it was going down well.

The effect of the wine on my empty stomach was predictable but welcome. The world, or at least Pau's restaurant, seemed a better place with each sip.

'UNA PALOMA BLANCA-A-A . . .' Pere's singing grew louder and the smells from his cooking more tempting, till at last the great moment arrived. A huge flat-bottomed earthenware *greixonera* emerged from the kitchen, held aloft by Señor Pau, who made a dramatic pause, then strode majestically to our table, vapour streaming from the bubbling dish.

The contents looked delicious – chunks of white fish floating in a mouth-watering stock of tomatoes, wine, olive oil, onions, garlic and herbs. This was his *Sopa de Pescado*, and from the first steaming spoonful, we knew it was no ordinary fish soup; and we needn't have worried about the freshness of the ingredients. This was a masterpiece, and we told him so.

A smug smile played at the corner of Pere's mouth. *'Ah sí'* . . . his fishermen *amigos* always saw him right too. He watched us drain our bowls, then replenished them to the brim with more overflowing ladlesful of his famous creation. By the end of this second bowlful, we felt so full that we wondered how on Earth we were going to cope with the next course. But some pieces of hake and monkfish still lingered in the dregs at the bottom of Pere's *greixonera*, and no amount of praise-laden protestations would persuade him that I just didn't have room for any more. Ellie was spared a third helping. She was a woman, Pere observed, and did not need so much sustenance as a man; but I must finish the *sopa*. It was good for me.

At that moment, the door flew open and a chill gust of air rushed in, followed by a crowd of noisy young men and their girlfriends, rubbing their cold hands and shouting good-natured jibes at Pere Pau while they packed themselves round the cramped tables. It was like a rowdy game of musical chairs. Naturally, there were two losers, and they gestured enquiringly if it might be OK to occupy the two empty places at our table. We gestured back that they were welcome. What else? We didn't really have much choice.

Pere ran the gauntlet of backslaps and banter as he dispensed bread, olives, beer and Coke throughout his full house. He was in his element – sparks flying from his cigarette with each insult that he mouthed at his boisterous guests. We had the impression that this little scene had been played many times before.

'Good, maybe he'll forget about our main course now that he has to feed this lot,' I said to Ellie. 'That fish soup of his was certainly fabulous, but after almost three bowls of it, I feel like a stuffed turkey.'

'No chance, mate,' chipped in the lad sitting opposite. 'Pere won't forget you; and if you've okayed his grub, it's a cert he gonna stuff you plenty more. That's his game with new punters. No danger, OK? Ya wouldn't coco.'

'Oh, I see. Well, I . . . I'm really looking forward to that,' I replied, slightly taken aback by the young fellow's fluent English, and trying not to look too distressed at the prospect of eating until I burst.

Our new dining companion was clearly keen to practise his English on us, possibly to impress the young lady sitting by his side, who gazed adoringly at him through big brown eyes, her chin resting on one hand while the other hand popped olives into her mouth with the steady rhythm of a robot. Revelling in this admiration, her beau redoubled the outpourings of his English.

We were in the company of members of the Andratx football team, he confided. They were celebrating that afternoon's victory over local archrivals, Calvià. He himself was *el striker* and had scored the decisive goal. Recognising the word 'goal', his *chica* squealed in delight and planted a big kiss on her hero's cheek. He pushed her away half-heartedly with that air of bored resignation which comes so easily to the bona fide sports star.

'I speak no bad English, eh?' The Andratx version of Maradona seemed to be making a statement of fact rather than asking a question, but we nodded in agreement anyway. 'I never done no English at school. Nah – I learn it all from the Brit holidaymakers in the beach bar where I work along in Magalluf. They learn you good, them guys . . . better than a fuckin' school teacher any day. No danger, mate.' He took a huge slug of beer from his bottle, then belched loudly in the face of his besotted groupie. His Brit pals had taught him well.

A deafening cheer went up, heralding the arrival of the team's first cauldron of *Sopa de Pescado*. Pere Pau shouted an apology in our direction. Our main course would be with us soon, he assured us, but he had to get the soup out to this bunch first. As we would probably know, the only way to keep pigs quiet was to feed them. A cacophony of grunting filled the room as the youngsters took the bait, and Pere returned to his cooking area, plainly satisfied that his performance was being well received.

'He treats us sumfin' shockin',' declared Maradona, waving his San Miguel bottle in the general direction of the kitchen, 'but he really loves us. All that pigs stuff – just an act – same every bloody time.'

'So you . . . you eat here quite often?' Ellie enquired hesitantly.

'Every Sunday night, darlin' – win or lose. This is like our clubhouse, see. Been like that for years. Look – the pictures on the wall. It's a bloody dump, innit? But the dinner only costs eight hundred pesetas each . . . three bloody quid. And the grub's good. Yeah, no danger, missus, the grub's well good.'

Ellie sipped her water, concluding that that little conversational interlude was over. I poured the last of the jug of wine into my glass, and *el striker* and his fan attacked their *sopa*. The steady slurping sound of team soup work reigned throughout the little restaurant for a minute or two, then the wonderful din of uninhibited Spanish table talk started up again.

'Hey, Pere, you miserable old bugger, how about some bleedin' service here!' shouted Maradona, looking over at us for signs of approval of his best English. But he was already getting all the approval he needed from his drooling *señorita*. How romantic this flowery foreign tongue must have sounded to her.

Pere made an exaggerated show of ignoring his other clients as he marched from the kitchen and placed two terracotta dishes before us. He closed his eyes, drew himself up to his full height and boomed theatrically, *'EL LOMO CON COL! . . . Bon profit, señores.'*

We were still visually marvelling at the succulent contents of the piping hot dishes when Pere returned to our table with another jug of wine and yet more crusty bread. He stood behind Ellie, arms folded and cigarette jutting erect from his mouth while we made our first incisions into this *gran especialidad de Mallorca*.

Tender young cabbage leaves had been used to envelop stuffed rolls of wafer-thin pork loin – each juicy package held together by a tiny wooden skewer. The bottom of each dish was smothered in a moat of rich brown gravy that sizzled around islands of baby potatoes, wild *seta* mushrooms, raisins and pine nuts. Whatever Pere Pau's other secret ingredients may have been, the appearance and smell of his country classic were irresistible.

And the taste? We didn't have to say anything. The expressions of ecstasy on our faces told Pere everything he wanted to know. He bowed in acknowledgement of our unspoken compliments and withdrew once more to his kitchen.

Meanwhile, over by the fridge where the beer was kept, our two young tablemates had joined a few of their chums in an impromptu action replay of the highlights of today's football game. A bread roll was being headed, back-heeled and kicked about on a pitch not much bigger than the door mat, while an endless uproar of jokes and laughter rose from the tables, and from a hidden corner of the kitchen, Pere's singing bellowed above it all . . . 'VALENCIA-A-A, RUM-PI-TUM-PI-YUM-PI-TUM . . .'

But Ellie and I were lost in his Lomo Con Col, growing more oblivious to the merry pandemonium that surrounded us with each fabulous forkful. In my own case, the Binissalem wine was another contributory factor towards this creeping oblivion. I had downed the second jugful without any assistance from Ellie, who had stuck sensibly with the mineral water throughout the entire protracted meal. I had now ceased even to bother about how I was managing to cram all that food and liquid into my body. Ellie groaned in disbelief as I mopped up the last traces of gravy with a large wedge of bread.

'This has been a super feast,' I slurred, 'and I'll never forget it.'

'You've made a complete pig of yourself,' said Ellie, 'and you'll probably be sick on the way home.'

The football team eventually started to leave, winding their scarves round their necks and disappearing into the windy darkness in high-spirited little groups. They'd all had a good day.

Maradona waved from the doorway. 'See ya, pal . . . Nice to meetcha, doll, and may the skin of yer arse never cover a tambourine. No bloody danger, eh?'

'Goodnight,' called Ellie, suitably charmed.

El striker strutted into the night, an arm draped nonchalantly over the shoulder of his clinging vine of a girlfriend, who now appeared even more awe-struck than ever by her idol's astonishing bilingualism. I just smiled the smile of the truly contented, and our young dinner partners were long gone by the time I was able to muster up the energy to wave goodbye.

The cats had started to drift back inside as the humans left, but it was a little dog that slipped in as Maradona made the final exit. Pere Pau chuckled and bent down to pick up the small brown waif whose decidedly curly tail was almost looping the loop with excitement. This was plainly a meeting of old friends.

'*Hola, Pepito, mi pequeño amigo!*' grinned Pere. Much head-patting, chin-tickling and face-licking followed, Pere's cigarette pointing skyward like a Roman candle as he strained to keep it out of reach of Pepito's busy little tongue. But Pere was fighting a losing battle, and rather than get rid of the essential cigarette, he elected to plop the little dog onto Ellie's lap. That suited Pepito just fine. He said his hellos to Ellie and snuggled down on her knee.

Sensing a growing atmosphere of bonhomie, the tortoiseshell cat crept purring from its bolt hole under Ellie's chair, climbed onto my outstretched legs, curled up and went to sleep.

I yawned and felt my head nodding, my eyelids growing heavy, my mouth falling open in readiness for a snore.

'Come on! Pull yourself together!' Ellie hissed. 'We're the only customers left. Señor Pau will want to close up for the night. It's time to go home!'

'Tranquilo, tranquilo, señores,' Pere called soothingly from the kitchen. There was no hurry, he assured us, and anyway, he had a surprise for his two very patient and – above all – discerning guests. He placed a large metal tray on the table. For Ellie, there was *un flan* – a glistening homemade creme caramel, a saucer of broken biscuits and some scraps of pork for Pepito, three cups of *café solo*, and a bulbous glass of *coñac* each for *el patrón* and myself.

At the sound of Pepito scoffing his little banquet at our feet, the tortoiseshell cat opened one eye, purred again, and went back to sleep. Clearly, the mouse population of Andratx had already been reduced considerably that evening.

Before sitting down, Pere trundled a mobile gas heater over to Ellie's side of the table. The *señora* had to keep warm on such a night. Find a cosy place indoors and stay put – that was the only thing to do when the *Tramuntana* was blowing. He pulled up a chair and raised his glass. *'Salud, señores!'*

It was already well past midnight, but Pere Pau wasn't ready to call it a day yet. He had his life story to tell us. I noticed Ellie politely stifling a yawn, but another *café solo* would cure her drowsiness, advised Pere. She must feel free to go to the kitchen and help herself – and not many were allowed that privilege, he stressed.

Even before consuming the two pitchers of wine, my slim grasp of Spanish had been stretched to the limit in order to understand Pere at times, but now, in my more 'relaxed' state, I found that I was catching his drift without too much difficulty. I jumped to the attractive conclusion that the consumption of such quantities of wine was clearly beneficial to one's comprehension of foreign languages – a theory subsequently wrecked by my wife when she pointed out to me that Pere's after-hours conversational Spanish had been spoken at a considerably slower pace and, for our further assistance, had been liberally sprinkled with English.

In any case, Pere Pau's story was interesting and it confirmed that he had the dedicated independence of the true eccentric. He had trained under several *gran maestros de cocina*, and he had himself served as head chef in some of the island's finest hotels. Why, he had even cooked for the King of Spain on many *ocasiones grandes* at Palma's Royal Yacht Club. But his heart had always remained in the country, and when he finally grew tired of trying to please fussy and often-unappreciative tourists, he'd gladly exchanged the five-star kitchens of Palma for this little place, his *own* little place, in his native Andratx.

Here he could recreate in his own *estilo especial* the traditional dishes of rural Mallorca, made from the wholesome local produce which had been used by his ancestors for centuries. He only attracted people who truly appreciated his food, he emphasised. (Why else would anyone come to his humble establishment?) And he didn't have to put up with those finicky *nuevos ricos* who were more interested in posing and complaining than in enjoying the great pleasures of good food. *'Ignorantes!'*

Pere grimaced at the thought and poured another two brandies. The *Tramuntana* howled outside. Lighting another cigarette, Pere looked down at the tortoiseshell cat sleeping on my legs, then at the other cats clustered snugly in front of the gas heater. He thought for a moment, then proclaimed, 'The public health officers know nothing – *absolutamente nada!*' How dare they suggest that his kitchen was unclean just because he allowed his cats in. Cats were clean creatures. How many public health officers could wipe their arses with their tongues? So leave cats out of it. Mice were the dirty ones. Mice spread disease – but there were no mice in his kitchen, thanks to the cats, of course. Cooks had encouraged cats in their kitchens for thousands of years before public health officers were even thought of. End of argument! Oh, and another thing – why should he not smoke while he was cooking? Smoke was created by fire, was it not, and no germs could live in fire. *Hombre*, if he cooked on a barbecue, there would be much more smoke *and* ashes in contact with the food. None of his cigarette ash landed on *his* food. 'Public health officers? *BASTARDOS!*'

Pere's fist thumped down on the table, causing the metal tray to take off and land with an ear-splitting clatter on the tiled floor. Startled cats fled in all directions, and the rudely-awakened Pepito leaped from Ellie's lap and landed squarely on the back of my tortoiseshell companion. I felt the now-familiar pain of cat claws sinking into my wounded legs.

Now it really was time to go home.

– TWO –

GAS & NUNS

'I know that there are no hens on your farm any more, because Francisca Ferrer strangled them all for the pot the day before you arrived,' smiled the old peasant woman, handing me a basket of large brown eggs. 'Feel – some of them are still warm,' she said, taking my hand and shoving it into the basket. *'Muy frescos, estos huevos, no?'*

They certainly were fresh – good old-fashioned fresh, with little, downy feathers stuck to the shells on smudges of hen droppings, just like the eggs I used to collect in my granny's henhouse when I was a kid. Suddenly, it made me realise how our more 'progressive' ways had made us accustomed to buying sanitised, supermarket eggs, all squeaky clean and officially graded in their boring little plastic boxes. Hell, I thought, there could be millions of kids these days who really believe that hens actually lay their eggs in six-packs.

I must have gone slightly dewy-eyed at the sight of those 'real' eggs.

'You do not like the eggs, *señor*?' asked the puzzled old woman.

'Oh, *sí, sí,*' I assured her. 'The droppings just reminded me of my granny for a moment.'

She scratched her grey head and muttered a string of unintelligible oaths, ending, I was sure, in *loco extranjero* – a common enough term in Mallorca, meaning 'crazy foreigner'.

Hastily thanking her for her kindness, I said that I hoped that I hadn't appeared ungrateful. It was difficult for me to explain in Spanish the feeling of nostalgia which her basket of honest, dung-dotted eggs had provoked, but I tried.

She laid a work-worn hand on mine, stopping me in mid-stammer. '*Tranquilo, señor* – do not worry. Spanish is not my native tongue either, and when I am forced to speak it, I must also search for the correct words. In the old days, we Mallorcans spoke only *mallorquín*, so now, if I have to converse in the tongue of the *españoles*, it is necessary for me to speak very slowly . . . just like you . . . like *un extranjero*, no?' Her eyes twinkled like little black pearls, and her face creased into a wide, open-mouthed smile, revealing five sparkling white teeth – two up, three down. It was one of those instantly infectious smiles, one of those visual tickles that defy resistance, and seeing my grinning reaction, the old woman clasped her hands together and chuckled quietly for a moment or two, her expression darkening into a scowl as she reverted to mumbling yet another stream of abuse to herself – this time about eggs, the Civil War and soldiers, I guessed, and the *profesión inmoral* of the *madres* of those thieving *españoles*.

I started to introduce myself.

'I know, I know,' she interrupted. 'Francisca Ferrer – *La Condesa*, the Countess, we call her – has told me all about you. *Com estàs?* I am Maria Bauzá, your neighbour in the next farm to the north. And I must thank you, *señor*, for the land which you have bought here separates my *finca* at last from that of *La Condesa* Ferrer – *gracias a Dios.*'

She had now made it fairly plain that she had precious little time for Francisca Ferrer (not to mention the *españoles*), but tempted as I was to ask the reasons why, I judged it best to contain my curiosity for the present, my prime objective as a newcomer to the valley – and a *loco extranjero* to boot – being to establish friendly terms with *all* of our neighbours . . . if at all possible.

We were standing under the sweeping, upward-curved outer branches of a sprawling fig tree near the stone wall that separated our two farms. Señora Bauzá's stooped, frail-looking frame was clad entirely in black, as is the custom of the older Mallorcan countrywomen, and I hadn't even been aware of her standing there in the shade as I walked through the orchard – the first of many times I would notice how the island's old *campesinos* merge so naturally into their tree-speckled landscape. In truth, if old Maria hadn't greeted me with a cheery *'Buenos!'*, I would have passed right by without even knowing she was there.

It had only just turned nine o'clock on a beautifully calm winter's morning, and the warmth of the early sun was already dispersing the gossamer-thin veil of mist that had been rising from the fields, still wet from the previous day's melted snow. My head was filled with the musty smell of the damp earth and the sharp, exotic fragrance of citrus blended with faint hints of juniper and wood smoke from the surrounding hillsides. A trace of wispy cloud hung over the table-top summit of *Ses Penyes* mountain, the rounded, rocky mass blocking the high northern end of the valley and sheltering the lower reaches from the worst of the cold blast which can hurtle down from northern Europe in winter, picking up speed and fury disguised as the Mistral as it funnels southwards through the Rhône valley in France, before howling unobstructed over the Mediterranean Sea to vent its final wrath on the Balearic Islands.

But the *Tramuntana* had blown itself out during the night, and a placid silence had returned to the valley, broken only by the distant barking of a dog on one of the high mountain *fincas*. Most of those remote farms are impossible to see from the valley below, the only sign of their existence being the thin straggles of white smoke from their chimneys drifting up through the densely wooded upper slopes on clear winter days.

'Very few people live up there any more,' said old Maria, following my gaze to the high ridges. 'But in the old days . . . well, there are little hidden valleys in the mountains up there where the soil was walled into terraces many centuries ago – they say by the Moors when they ruled this island. And families used to scratch a living from those narrow *bancales*, growing whatever they could. *Sí*, and in the woods, they would rear a few pigs – little, black Mallorcan *porcs*. Ah, those *porcs* . . . what a taste, and all because of the acorns and things. The pigs had to find their food by rooting among the trees, you understand. Pigs and forests go well together, no? *Sí*, the forest floor feeds the pigs, and the pigs keep the forest floor clean. *Un sistema perfecto, eh? Sí, sí* – no forest fires in the old days . . .'

I started to ask how they got water up to those mountain farms before the days of tanker trucks, but her train of thought was not to be diverted.

'You need a pig, *señor*. Everyone with a fruit farm needs a pig . . . maybe two. Depends on how much fallen fruit you have. You need to make use of everything on these farms, and pigs are good for that. Hens too. You need new hens now that *La Condesa* has wrung the necks of the old ones. Yeugh!'

She shook her head rapidly and shuddered, as if she had just tasted something foul. Francisca Ferrer was definitely not flavour of the month.

I thanked her for her useful advice and assured her that I would see to buying a pig and some hens once we had settled in. But I was more urgently concerned about our two hundred or more trees of oranges that were almost ready to pick. Perhaps she could suggest a merchant who might buy the fruit from me?

'How did they get the water up there? *Hombre*, they did not . . . they could not,' she shrugged, reverting to my last-but-one question with no sign of a mental gear-change. 'When it rained, all the roof water ran into an underground tank – *un aljibe*, just like the one at your house, no? *Sí*, and not many mountain farms have wells like down here in the valley, so the winter rain in the *aljibes* had to last most of them all summer long.' She rubbed her forefinger and thumb together. 'Water is gold to the Mallorcan farmer, eh, and many feuds have been fought over wells and their precious *agua*. You will learn soon enough, *señor*. Ah *sí*, you will learn. *Agua, agua, agua* . . .'

She nodded sagely and shuddered again, prompting me to wonder if the Señoras Bauzá and Ferrer were adversaries in a water war. The words Ferrer and *agua* certainly seemed to have the same nauseous effect on the old woman.

'Tomàs Ferrer told me that the well on this farm is one of the best in the valley, so we're lucky. Of course, the Ferrers still have sole use of the well at weekends. That was the deal when we bought the place, but my lawyer did assure me that this sort of arrangement is quite common here when part of a farm is sold and part is retained by the seller. It seems a fair enough system . . . if both parties play the game. You must be very trusting and co-operative people, you Mallorcans, eh?' I quipped.

Old Maria glanced quickly at the orange trees. 'Just ask Jaume, my son-in-law, about selling your fruit,' she advised, picking up again on my penultimate topic. 'He helps me to work my *finca*, you see. I cannot lift the fruit crates any more – not like I could in the old days.' She looked down and tapped the fingers of her left hand in turn. '*Sí*, I am . . . ehm, eighty-two, you know, so Jaume deals with selling the fruit. But I still look after the money, though. I am not too old for that.' She beamed her five-tooth smile again. '*Hombre*, I will never be too old for that. *Nunca jamas!*' Laughing quietly to herself, she turned to leave. '*Adéu, señor*. My regards to your wife . . . and I hope you enjoy the eggs.'

The old woman hobbled off through a colonnade of lemon trees, paused, half-turned towards me, and shouted out in a shrill treble which could have been heard throughout the whole valley, 'AND DO NOT TALK TO ME ABOUT SHARING WATER WELLS WITH NEIGHBOURS!' She winked impishly, then added in a loud whisper, 'You will learn soon enough, *señor*. You will learn.'

Hell's bells, I really did have a lot to learn, I thought to myself while I trudged back through the fields to the house. God, look at all those bloody fruit trees . . . I still had to find out how to feed them, prune them, spray them, irrigate them, and the only ones I could even identify yet were the ones with oranges and lemons hanging from them. Still, at least old Maria Bauzá had given me a bit of a lead regarding the selling of the fruit. That was good . . . but now there was this tricky water-sharing business, and pigs, and hens, and all the weeds growing under the trees . . . and I didn't even have a tractor yet, and . . .

'Have you had a shower this morning?' Ellie called from the bathroom window as I emerged from a little grove of almond trees in front of the house.

'No, not yet – but I didn't think you'd be able to smell me from that distance.'

'Never mind that. There's a problem with the shower. No hot water.'

The boiler was situated in a dark corner of the *almacén*, a large room used for sorting and storing the fruit, which took up more than half of the ground floor of the house, and by the time I arrived on site, Ellie was already there, banging the gas cylinder on the floor and tugging at its rubber connecting pipe.

'Right,' she said, 'there's gas in the bottle, and it's joined up to the boiler OK. The pilot light's still on, so that's all right. But wait a minute. Mm-mm, there's a smell of gas, isn't there? Yes, it's a job for a plumber.'

There was nothing left for me to say, except a rather futile, 'Yeah, and if you ask me, that old boiler looks a bit small for the size of the house, anyway . . .'

Juan the plumber was no ordinary plumber. He was a plumber *and* an electrician, so his tiny front shop in the Andratx market place was an Aladdin's cave of mini-chandeliers, wall lights, electric irons, toasters, radiators, electric fans – even a couple of fridges, a washing machine and a microwave oven. The shop was tended by Juan's wife, a petite, happy-faced girl who looked no older than a teenager herself, yet she had three tiny tots pulling at her skirts behind the counter, and we could hear the sound of at least one more baby crying in the back room.

'My husband is very busy,' she told us.

'Not busy enough, it seems,' murmured Ellie, looking over the counter at her swarm of under-fives.

'Nevertheless, he will be at your house to fix your water boiler tonight at eight o'clock. We live not far from you, so he can call by before he comes home for dinner.'

'So much for the so-called *Mañana* Syndrome,' I remarked as we left the shop. 'That's what I call service.'

The trucks which delivered the butane gas bottles were too long to turn at our gate, so Francisca Ferrer had told us to leave our empty canisters (with the refill money underneath) at the end of the lane on Wednesdays, although – in emergencies – we could also get a full canister on Mondays and Fridays by driving around the streets of Andratx until we came across one of the trucks making its town deliveries. This being a Monday, we had put the empty canister from the cooker into the car, and by timing our truck search to start at 10 am prompt, we guessed that we had a fair chance of hitting the jackpot immediately by heading straight for the main square, the Plaza de España. And there, as expected, was the *butano* man, relaxing over his mid-morning coffee and *coñac* with a few local worthies outside the Bar Nuevo, his orange-coloured *camión* fully loaded and totally abandoned, half mounted on the pavement at the corner of the square.

'*Perdón,*' I said, when at length he had decided that it was about time to saunter back to his truck, smoke streaming from the fat, stubby cigar which wobbled between his lips like a broken brown banana in a face-shaped strawberry blancmange. '*Perdón*, but I'd like to buy a bottle of gas, *por favor.*'

He heaved a full canister from the truck and replaced it with my empty one. I handed him some notes, and he wedged his lit cigar into the neck of my full canister while he fumbled for change in the pockets of his overalls. I closed my eyes and waited for the bang, but the only explosion was a crepitus chuckle from the *butano* man.

'Isn't that a bit dangerous – smoking when you're working with gas bottles?' I asked edgily.

'Dangerous?' He handed me my change and raised his shoulders. '*Coño*, if someone crashes into me on the road – and there are always plenty of *lunáticos* driving around on this island – this *camión* will go up like an atom bomb, and me with it. I live on borrowed time, *amigo*, so I enjoy myself while I am still alive . . . I smoke.' He retrieved his cigar and climbed into the truck.

'While you're there, why not buy two full canisters,' Ellie shouted from the car. 'We could do with a spare.'

'*No es posible, señora,*' the *butano* man called back, shaking his head with one eye closed against the rising smoke from his cigar, now stuffed safely back in his face. In order to buy a second full *botella*, he explained, we either had to give him a second empty *botella* or show him a certificate from the *butano* office verifying that we were entitled to buy another full *botella*. An empty *botella* or an official *certificado* – and that was that. 'I am sorry, *amigos. Es el sistema*.' He raised his shoulders again, revved up the engine, and rumbled off, leaving puffs of cigar and diesel smoke drifting up over the square like the farewell signals of the last of the Mohicans.

'Why the blazes do we need a piece of paper to allow us to buy a spare bottle of gas?' I fumed. 'Why can't they just charge a deposit on the bloody thing like they do at home?'

'Like the man said – it's the system. There must be a kind of logic in there somewhere,' Ellie concluded, 'but we won't get to the bottom of it standing here. Let's go to the *butano* office.'

'We would like *un certificado* to allow us to buy *una botella de butano*. We have no empty *botella*, so we need *un certificado, sí?*' I droned parrot-fashion to the disinterested *butano* clerk. A coach roared past on the busy road outside.

'*Qué! Un certificado?*' The clerk frowned and looked at me as disbelievingly as if I had just broken the news to him that his wife had given birth to sextuplets, every one a spitting image of the Pope. '*Por qué? No comprendo.*' Another heavy vehicle thundered past.

Suddenly, my inadequate Spanish seemed totally useless. I didn't really understand why I had to ask for a certificate, anyway, and every stilted attempt I made to elucidate was being drowned out by the din of passing traffic. I was sorely tempted to employ the old Brit-abroad tactic of bawling loudly and slowly at the clerk in English, but I instinctively knew that this would be pointless. This guy could see that I was sweating blood, and the bastard was enjoying every minute of it. I struggled on in Spanish.

After several more abortive attempts to explain what we wanted, I was about to suggest to Ellie that the simplest solution would be to go all-electric, when the clerk pulled a self-satisfied smirk and announced in English that he couldn't issue the certificate we wanted unless we presented him with a receipt for the new appliance for which the extra gas canister was required.

'But we don't *need* another gas appliance,' Ellie objected, her normally inexhaustible patience rapidly running out, 'just a spare bottle. I mean, all this certificate, appliance, receipt business – it's just plain crazy!'

The clerk leaned back in his chair, crossed his feet on his desk, and declared smugly, 'Maybe crazy, *señora*, but the receep system she works good. And nobody never steal no empties in Spain.'

'That situation may well change today, mate,' I mumbled, taking Ellie by the arm and leading her towards the door before she could clobber the clerk with her already-cocked handbag.

She was still in a rage when we got back to the house.

'Connect up the new canister,' she barked, 'and I'll fix something for lunch with old Maria's eggs. I'll imagine it's that damned clerk's brains I'm scrambling!'

I snapped the cooker's hose connection onto the top of the full cylinder, then tried to turn on one of the gas rings. The knob came away in my hand.

'What the hell next, Ellie? Now the bloody cooker's falling to bits!'

'Don't dramatise. Just turn it on with a spanner or something. Improvise.'

I duly grabbed the knobless spindle with a pair of pliers, and gave it a twist. The hiss and smell affirmed that the gas was getting through all right, but solving that problem had merely created another. I couldn't turn the gas off again. The spindle was now firmly stuck in the 'On' position, and no amount of wrenching or swearing would budge it. In desperation, I gave the spindle a hefty thump with the pliers, thus producing problem number three – the spindle disappeared with a clatter inside the cooker, and the gas continued to pour out.

'We're really in trouble now,' I spluttered, staggering back from the cooker. 'RUN!'

'Use your loaf, for God's sake. Disconnect the canister again before you either gas us or blow up the house!' She clashed the eggs back in the fridge and slammed the door shut. 'OK, Red Adair – yesterday we had a cooker but no gas – now we've got gas but no cooker. Any bright ideas?'

* * * * * * * * *

The Bay of Palma had never looked more glorious. We drove down off the *autopista* from Andratx to be greeted by the midday sun, a golden ball suspended against a curtain of sapphire blue, shining down on the aquamarine expanse of the Mediterranean Sea on which sequins of reflected light glittered like diamonds. The recent wash of melted snow had left the landscape refreshed and renewed, the green wooded hills surrounding the beige, circular bulk of Bellver Castle away to our left mirroring the clear sunlight in an emerald glow, and offering up subtle scents of pine and myrtle to blend with the salty tang of the sea in an invigorating cocktail of balmy winter air.

We passed the escalating terraces of Palma's most chic hotels and restaurants which boast (and charge sweetly for) unrivalled views of the bay from those densely urbanised slopes where once only sheep and goats grazed peacefully in millennial groves of olive and almond. We continued on along the Paseo Marítimo, a spectacular sweep of palm-lined highway which hugs the curve of the bay on a broad garland of land reclaimed from the sea. It serves, not only as a vital artery for the transportation demands of mass tourism, but also as an elegant avenue. Native Palmesanos and visitors alike stroll or linger for refreshment at a promenade café overlooking Palma harbour, once both a haven for marauding corsairs and the home port of honest sailing ships that plied the island's trade on the oceans of the world.

I stopped at traffic lights and took time to look out at the old quayside. Fishermen were drying and mending their nets in the sun, while nearby a few tourists sat sipping drinks beneath the tall date palms and gazed up at the city's historic landmarks protruding above the now redundant impregnability of the old sea walls; the ancient windmills of Es Jonquet, now transformed into trendy bars and nightclubs; the handsome mansions of rich merchants of a bygone maritime era, their stately homes now sub-divided into apartments of great style for today's successful traders in tourism; and finally, the awe-inspiring magnificence of Palma's mighty cathedral, La Seo, the honey-gold limestone of its twin spires and pinnacled rows of flying buttresses glowing with a rosy hue in the winter sunlight, and manifesting the eternal domination of the church over the city and the sea.

The lights changed to green, and in the couple of seconds it took me to engage first gear and release the clutch, a dozen car horns were blaring impatiently behind us.

'TURISTAS!' yelled a taxi driver, bouncing the palm of his hand off his forehead and glaring furiously at me as he roared past in a mad dash to gain pole position at the next set of lights. I still hadn't mastered the race track technique which must be exercised in Palma if you want to go anywhere by car without the native drivers making you feel about as welcome on their streets as a pork pie salesman in Tel Aviv.

The urban racers nose-to-tailed it past us (on both sides), mandatory cigarettes dangling from heavily moustached lips, left hands draped out of side windows, and horns honking at the merest suspicion of the driver in front easing his foot off the gas pedal by even a millimetre. And the presence of any eye-catching female driver only added to the chaos and considerably increased the risk of a multiple pile-up, the lady being the target for torrid ogling from all sides – some particularly lascivious males even hanging out of their drivers' windows to look back and eye her car up and down as if it were an extension of her own body . . . and all the while belting along flat-out almost glued to the car ahead.

Armed with the cooker's instruction manual, we were heading in the general direction of the agent's showroom in Palma's busiest shopping district.

Wheeling left off the Paseo, we hurtled along in a three-abreast charge past the lush gardens of the old Moorish Almudaina Palace, where cool fountains played amid sub-tropical arbours and whispered their soothing song to a line of weary-headed horses, standing on three legs and hitched to open-topped carriages in which their drivers sprawled and dozed while awaiting a rare winter fare from the occasional passing group of sightseers.

Without warning, the avenue funnelled into a narrow lane beneath the arcaded canopy of plane trees flanking the broad boulevard of the Borne, the dodgem derby grinding to an impatient, Indian file crawl past the lavishly dressed windows of up-market shops selling high-fashion clothes, luxury leatherware and expensive watches and jewellery below magnificent, sculpted stone facades of grand and ornately balconied Spanish elegance.

'The cooker place is up that street on the left, just ahead,' Ellie said, 'so keep your eyes peeled for a parking space.'

'You'll be lucky. There's no left turn, and anyway, there's about as much chance of finding a parking place around here as there is of Pere Pau's restaurant being included in the Michelin Guide.'

Picking up speed and lanes as the street fanned into the Plaza Rei Joan Carles I, we were swept along again in the free-for-all stampede, veering right at the Bar Bosch, where the pavement tables were crowded, as ever, with rubbernecked tourists in shorts and flip-flops, dapper businessmen sitting cross-legged in cashmere suits and Carrera shades, and see-and-be-seen students eyeing up the talent for as long as is humanly possible without actually buying a drink.

'Say goodbye to the cooker shop,' I grumped. 'We'll be stuck in this traffic jam for the rest of the day, and God knows where we'll end up.'

'Look!' Ellie shrieked. 'There's an underground car park right there, and there's a green *libre* sign, so there must be a space. Go for it!'

She gripped the sides of her seat as I swung the car to the right, nipping in between two buses and almost colliding with a carload of nuns who were signalling to go into the car park too. It was one of those fifty-fifty situations, and back in Britain I would surely have shown good manners by letting the ladies go ahead, but this was Spain, I had just been subjected to a scrotum-tightening initiation into their driving methods, and besides, I was only a *loco extranjero*.

'Bugger it! When in Rome!' I snarled, swerving brazenly in front of the nuns, and just beating them to the ticket barrier.

'Peter! That was awful,' gasped Ellie, staring at me aghast. 'I'm absolutely mortified. Those poor nuns. Your 'When in Rome' attitude is right out of order when it comes to being unchivalrous to women in Spain – especially nuns.'

'So it's all right to letch after women like those randy Spanish drivers do, but nicking their parking space is out of order. Is that what you're saying?'

'Something like that, and I'm completely ashamed of you. That wasn't nice, and if those nuns weren't above such things, they'd be giving you a good tongue lashing. It's a good job for you that they're gentle ladies of the cloth.'

'Could be, but a tongue lashing from a nun would still be a cheap price to pay for a parking place in this area. I got in first – fair and square. The nuns can look after themselves.'

'Ah, *buenos días*. And how may I be of service to your honours?' smarmed the salesman, gliding over the showroom floor on patent leather pumps that peeped from under his circa 1970 flares. A podgy little man with unnaturally black hair slicked down over his skull like a greased bombazine swimming cap, Zapata moustache trimmed slightly higher at the left to focus attention on a gold eye tooth which he exposed in a permanent one-sided leer, he had the appearance of a swarthy Peter Lorre gigolo with Bell's palsy, and looked as though he might have been more at home tangoing the night away in some dingy, downtown dancehall than flogging fridges and cookers in a swank, uptown department store.

I handed him the cooker manual, pointing my finger at an illustration of the control panel. '*Roto* . . . the knob, the spindle . . . all *roto* . . . all broken.'

The leer equalised into a full-face grin, his chameleonic eyes sparkling in commission-anticipating delight. Our particular cooker was obsolete, he elatedly informed us, the manufacturer having gone bust years ago. No spare parts were still available, of course, and in its present condition, the cooker was positively dangerous – *muy peligroso . . . una bomba!*

Chasséing over to the display area, he begged to advise our honours that the only solution would be to scrap the old cooker forthwith, and buy a new one . . . and in this showroom, we had the good fortune of being able to choose from the most comprehensive range in Mallorca.

Five minutes later, we were several hundred pounds lighter and the proud (if somewhat stunned and disgruntled) owners of a sturdy, wide-bodied model with a special compartment alongside the oven in which to conceal the gas canister. *Muy importante*, that.

A wise choice – a cooker well suited to the traditional farmhouse kitchen, confirmed the salesman, handing Ellie the guarantee and receipt. '*Muy importante*, the receipt . . . to enable your honours to purchase the new *botella de butano* for the special compartment. *Muy importante*, the receipt,' he stressed.

Didn't we know it! And bidding farewell to *muchas pesetas* right out of the blue put me in no mood for the sight that greeted us back in the underground car park either.

'I don't bloody well believe it,' I wailed as my eyes grew accustomed to the gloom. 'We've got a flat tyre!'

Ellie stood humming quietly to herself while I rummaged angrily in the boot for the spare wheel.

'Now, where's the hole to slide the jack into?' I was down on my knees now, peering under the car, fumbling in the darkness for the mounting, and getting absolutely nowhere.

Then Ellie opened the car door. 'There you are. That's turned on the interior light. Maybe that'll help.'

'Ah, that's better. There!' I pushed the jack firmly home. 'Now we're getting somewhere. Good idea that, Ellie.'

I quickly pulled my head out from under the car, only to hit it a resounding whack on the underside of the open door. The pain was excruciating. 'What the bloody hell is that still open for?' I yelled, lashing out a sideways kick at the flat tyre, missing, and battering my cat-lacerated shin against the corner of the bumper.

'Jesus Christ!' I howled in agony. 'JESUS H. CHRIST!'

Spot on cue, the nunmobile materialised out of the shadows, slipping slowly past en route for the exit. I could just make out the serenely satisfied expressions on the sisters' faces as they looked down at me sitting pathetically on the concrete floor nursing my head and leg.

'They couldn't have. They wouldn't have,' I thought aloud.

An elderly nun smiled out of the rear window of their car, and made a bi-digital gesture at me that only the most devoutly gullible could have mistaken for the sign of benediction.

'They could have, and they most certainly should have,' Ellie opined dryly. 'But as I said before, I'm sure those ladies are above such things. Your flat tyre is more likely a case of divine retribution.'

I wasn't so convinced.

Back at the *butano* office, we presented our new receipt to the clerk who then made out the much-valued *certificado* with a told-you-so smirk.

'There you go, *señores*. Like I say, the receep system she is very, very good, no? Now you gonna have two gas *botellas* for only one cooker. *Olé!*'

We conceded defeat silently and went home.

* * * * * * * * *

Despite my misplaced faith in the improbable, Juan the plumber (like plumbers everywhere) eventually turned up three days later, our increasingly frantic calls to his wife having been met with all the stock excuses – a burst pipe here, a choked drain there, and every job an absolute emergency that had to be attended to *inmediatamente*. Why, she hadn't seen him home before midnight for almost a week.

'*Es un desastre!*' Juan declared, looking extremely fresh and bright-eyed for such an over-worked and under-slept young man. 'Your water boiler is useless. It is leaking gas. *Es una bomba!*'

So now we had another potential bomb disguised as one of Señora Ferrer's modern conveniences. Bloody marvellous! First the cooker, now this. I felt myself going all light-walleted again.

He knew this boiler well, Juan informed us. It had been dangerous for ages, but Ferrer was too mean to replace it. It had always been too small to provide hot water for a house this size, anyway. He pointed to the little porcelain tub underneath it. That was what the boiler had been designed to provide hot water for – one sink!

'Yes, I thought it was a bit too small as well,' I agreed feebly, realising that young Juan must have been regarding us as the typical, starry-eyed foreign idiots who had been an easy mark for being sold a houseful of rubbish by a pair of crafty locals.

'And by the way, how is your *electricidad?*' enquired Juan, donning his electrician's hat and squinting suspiciously at the ancient fuse box on the wall. Hadn't we had any problems yet?

'Well, now you mention it, the fuses sometimes blow if we try to use the electric kettle and the toaster at the same time,' I admitted, feeling increasingly stupid.

Juan shook his head and shrugged a slow Spanish shrug. *'Hombre, es un catástrofe.* You have *problemas . . . problemas grandes.'* What this house needed was a complete electrical overhaul, he announced, looking me squarely in the eye. It was a miracle that there hadn't already been an electrical fire or a gas explosion . . . or both. We would need a much bigger fuse box, *un interruptor moderno*, to comply with regulations, and much of the wiring would have to be replaced with the correct gauge, the present stuff being only heavy enough for a few lights. *'Hombre – un catástrofe potencial!'* Then there was the problem of the water boiler. We should forget about silly little contraptions run off a gas canister. Get an electric boiler – a good big one – much cleaner and safer. Keep the gas bottle as a spare – always handy, a spare *botella de gas*, no?

Tell me about it. If only he had known what a wild gas chase we had already been on to get just one spare canister, and now we were going to have two!

'Ehm, could you have a look at the washing machine while you're here?' Ellie asked timidly, dreading the inevitable reply. 'It . . . well, it doesn't seem to spin very well.'

'Washing machine? *Señora*, I must tell you honestly . . .' Juan paused and ambled over to the vintage *lavadora*. 'This can of bolts has been useless for years. The fast-spin function broke down nearly ten tears ago, and old Ferrer would not spend the money to have it fixed. His wife has done the washing in their Palma apartment ever since. This heap of junk should have been thrown out long ago. *Basura!'* He gave the base of the machine a gentle kick, there was a metallic rattle in its bowels – then the door fell off. Juan did his best to conceal a smile by stroking the end of his nose with his thumb. '*Lo siento, señores*. I am sorry.'

Ellie and I looked at each other and started to laugh – rueful, shoulder-shaking giggles of pathos at first, then loud, almost-hysterical laughter – tear-wiping, thigh-slapping, leg-crossing laughter. It was now as obvious as a dog's balls that we had been well and truly taken for a couple of mugs by the Ferrers, and laughter was the best medicine. *Hombre*, it was the only medicine!

Juan moved over to the door, toying nervously with the handle, and probably thinking that the old house was now occupied by a right pair of foreign nutcases. He was ready to leave – quickly, if necessary.

'Don't worry, Juan,' I said, struggling to regain my composure. 'It's just our way of relieving the, ehm . . . *problemas*.' I gave him a reassuring nudge. 'Know what I mean?'

I was quite sure from the look of confused mistrust on his face that he most certainly did not know what I meant, so I prudently got back to the business in hand. 'OK, Juan – you make a list of all the things that need to be done, work out an estimate of the cost, and we can pick it up from your shop tomorrow. And if the price is right . . .' I placed a hand on his shoulder, and he drew away as if he had been touched by a live cable. 'And if the price is right, and if you can promise a quick job, you'll have a deal. OK, Juan?'

'OK, *señor* . . . *muy bien*, OK,' he stammered, fumbling for the door handle. '*Mañana* . . . OK.'

The door swung open and he was gone.

'We've scared the living daylights out of the poor chap,' said Ellie, still dabbing her eyes. 'He'll never set foot in this house again.'

'Don't you believe it. We're talking about a sizeable contract here, so he'll be keen to get the work, even if he does think we're escapees from an asylum for *loco extranjeros*. You'll see.'

This time, my hunch was right, and Juan proved the internationalism of plumber-profit-motivation. His estimate was ready for us at his shop the next day, and although (true to the trait of his trade) his price was roughly double what we expected, we were desperate, so we told him the job was his – but only if he could start *inmediatamente*.

'*Inmediatamente, señores!* I will pick up a new boiler and I will install it today. It will be a temporary system, of course, until the new *interruptor* and wiring are in, but it will give you abundant hot water – *inmediatamente!*'

While Juan was still high with the elation of having secured our contract, Ellie grabbed her chance to squeeze a bargain price out of him for the new washing machine which was on display in his shop.

'*Sí, una lavadora excelente, señora. Una máquina fabulosa!*' Juan enthused, tapping the base of the gleaming new washing machine lightly with his foot. 'Good for relieving the, ehm . . . *problemas*, no?' He looked at us apprehensively, gave me a dig in the ribs with his elbow, and let out a peal of forced, nervous laughter.

We joined in obligingly, no doubt confirming in the plumber's confused mind that we were a couple of perverted weirdos who got their jollies by watching someone sinking the boot into washing machines.

Having decided, perhaps, that it was best not to cross such oddballs, Juan stuck to his word, and by that same evening we had copious quantities of steaming hot water bubbling through the pipes from a shiny, white electric boiler of grand proportions. After days of taking birdbaths with kettle-heated water in the washbasin, we revelled in this rediscovered luxury – Ellie in the upstairs bath, and I in the downstairs shower. The rafters fairly rang with singing and splashing. And yes, I told myself, the old house *did* have a happy atmosphere after all.

Ah, the magic of hot water, and oh, how easily it lulls the bathing human into a false sense of wellbeing and goodwill. But, for the moment, we only knew that life had taken a definite turn for the better, and even the lumpy old bed seemed more comfortable when we snuggled under the covers that night, feeling all pink and cosy and clean.

'Incidentally, Ellie,' I said, switching off the bedside light, 'I picked up that wheel from the garage today, and there hadn't been a puncture after all. All they had to do was reinflate the tyre.'

'Mm-mm . . .'

'So you see, even if, as you thought, it *was* heavenly justice that was done in that underground car park, it seems to have been administered by an earthly hand.'

Ellie was already asleep.

–THREE–

SHOTGUN SUNDAY

Sunday is the day of the Mallorcan family, the day for a drive into the mountains to cook a *paella* over a wood fire, the day to go to a football game or bullfight, the day to visit a quiet, pine-fringed cove for a stroll by the harbour or, perhaps, to go on a fishing trip in the family boat, but above all, Sunday is the day for crowding into one of the many country restaurants that specialise in catering for that gloriously clamorous afternoon celebration of prandial conviviality – the Mallorcan Sunday Lunch.

Sunday had become our favourite day of the week.

'Look at that sky, Ellie,' I shouted, leaning out of the bedroom window and breathing in great lungfuls of the scented air. 'Look at that deep blue against the pale green eucalyptus leaves. It's the week before Christmas, don't forget, and it's like a perfect summer's morning back home. In fact, I've never seen a summer's morning like this at home. And just smell those smells . . . the oranges, the lemon blossom, the mountains. Aahhh, it's fabulous!'

'Shut up. You sound like a holiday brochure. Anyway, it's the middle of the night and I'm still asleep. Go away.' She rolled over and pulled the sheets over her head.

I had been up for hours and had already breakfasted on three of old Maria's brown eggs. I'd forgotten that eggs could look and taste anything like that. Golden yolks to rival the colour of ripe oranges in the red glow of evening, and a taste that could only be described as second to none – the peerless product of a clucking lineage which must have spanned centuries of unhurried scraping for seeds and grains in the sun-warmed soil, and nibbling contentedly at all the exotic chicken feed which must abound on a little Mallorcan *finca* like Maria's.

Ellie had never been happy about being disturbed too early on a Sunday morning. It *was* the day of rest, after all. That was her belief, and her aim, therefore, was to stay snugly in bed for as long as possible.

I'd never objected, but this was different. This wasn't a typical Sunday morning in Britain where everything seemed as cold and dismal and grey as the sky's covering quilt of rain clouds. This was Mallorca, the queen of the Mediterranean, an island of the gods, overflowing with the gifts of nature, and the sun was already high in a cloudless sky.

I was even beginning to think like a holiday brochure, but I didn't care. I meant it.

'Ellie, it's nearly eleven o'clock, and if we don't make a move now, we'll have no chance of getting into any of the good places for lunch. But suit yourself – we can fix up a bite to eat here on our own, if you like, or we can get right into the Mallorcan Sunday spirit and pile into a lively eatery like the natives.'

The ultimatum had the anticipated effect. Ellie rose dream-like from her bed and floated – eyes still closed – into the bathroom in a swirl of dressing gown.

'Consider me one of the natives,' she mumbled.

'*HOLA! HEY, AMIGO! I AM RAFAEL . . . ORANGES . . . NARANJAS!*' came a call from the lane. It was old Rafael, our first customer. He had started to come up from the village for a bag or two of pick-your-own oranges on most Sunday mornings, usually with a few of the goats which he kept in the miniature shanty town of patched-up shacks and tumble-down sheds in the little yard behind his terraced cottage on the main street. Any time he had to go out, his habit was to take three or four goats along to pick up some free grazing at the roadsides – and today was no exception. I could smell the goats before I reached the gate. Rafael was peering in through the metal bars, two of his nanny goats and their kids already tethered to fence posts at the other side of the lane. They were munching away busily at the weeds, but when they heard me turning the lock, they raised their heads, looked over at me inquisitively and bleated in unison.

'*Buenos días*, ladies,' I smiled, returning their greeting. 'Oh, and *buenos días* to you too, Rafael.'

He was a small, stocky man in his seventies, with a round, cheery face half-hidden under an old corduroy cap. His clothes were due for a long service medal and they positively reeked of goats, which didn't seem to bother the little boy who was clinging shyly to the old man's baggy trouser legs while doing his best to hide from me.

'*Mi nieto,*' said Rafael proudly, patting the little lad on the head.

'Your grandson, eh? He's a fine boy. Your first grandchild?'

Rafael looked offended. 'No, *amigo*, I have fifteen,' he stated indignantly. 'Four sons, three daughters and fifteen grandchildren . . . with two more up the pipe.' He nudged me with his elbow and winked. 'We are good breeders, we e'*pañoles*, no?'

I gave him a little slap on the back. 'OK, Rafael. Go and pick your oranges and I'll see you back at the house.'

Given any encouragement, Rafael would probably have stood and talked all day – or at least until his goats had stripped the verge of all vegetation, but although he was a likeable old character whose visits I always enjoyed, I still found it a strain to carry on a conversation with him. He was from Andalusia in southern Spain and, for a foreign rookie like me, at times his accent was virtually impossible to understand. I could just about cope with the mainland Spanish way of lisping the letters 'c' and 'z' when the rules of the language demanded, but Rafael also had the Andalusian habit of dropping the letter 's' from his words, and those disappearing sibilants were a killer. Slowly-spoken pleasantries were one thing, but if he got excited about a subject and slipped into full-speed jabber, he might just as well have been speaking Martian as far as I was concerned.

Walking back round to the front of the house, I could hear the shrill voice of our weekend neighbour, Francisca Ferrer. She and Ellie were exchanging their own brand of shorthand dialogue as usual, but this time I could detect a note of aggro. Francisca had evidently come over with a bundle of pet food for us to give to her dogs and cats through the coming week, but she was clearly not a happy woman.

Why were her dear little Robin and Marian not being allowed to sleep in our kitchen during the week, she wanted to know? That was the agreement we had made when we moved in, was it not? Now she found that the dogs' beds had been moved into a shed at her *casita!* She had fully intended to come over and confront us about this when she arrived from Palma on Friday evening, but she had been far too upset, had been stricken by a migraine attack and had been put to bed by Tomàs. *'Madre de Dios!'*

'Hmm. Well, I've been meaning to explain,' Ellie admitted half apologetically, 'but frankly, I was just too embarrassed.'

'Cómo?'

Señora Ferrer looked blankly at me, hoping for a morsel of linguistic assistance. She didn't get any. I was leaving this one to Ellie.

'Your dogs — Robin and Marian — your *perros* are not really house-trained, are they? That's the problem — *the problema, sí?* The *problema* is that they're . . . oh, hell's bells!' Ellie was becoming uncharacteristically agitated. 'What's the Spanish word for dog mess?'

Francisca's face was a picture of confusion. *'Cómo? No comprendo.'*

Ellie sighed deeply. She walked deliberately over to the house and, pointing at the bathroom window, she shouted back to the hapless Francisca, 'See . . . bathroom . . . dogs do bathroom, *sí?'* Then, pointing towards the kitchen window, she added, 'Dogs, your *perros*, do bathroom in kitchen — the *cocina*. No good!'

Señora Ferrer shook her head sadly, the corners of her mouth drooping, her chin quivering. 'Why do you want Robin and Marian to sleep in the bathroom?' she enquired tearfully. *'Qué cruel!'*

'No, no, no!' called Ellie, waving her hands frantically at our bewildered neighbour. 'No sleep in bathroom . . . crappo in the *cocina*! Get it?'

No such luck. The language barrier was holding firm.

I had always regarded my wife as the epitome of good taste and ladylike reserve, so her ensuing performance came as something of a shock. A loud, lingering raspberry exploded from her lips, followed by a booming 'WOOF! WOOF!' With her face contorted in an expression of pseudo disgust, she looked at the upturned sole of her slipper, held her nose, stared directly at Señora Ferrer and mouthed an exaggerated 'POO-OOH! . . . YEE-EE-EUGH!'

Old Rafael had heard some of this bizarre commotion as he approached the house on his way back from the orange grove. Perhaps fearful of the consequences of disturbing some form of Scottish pagan ritual, he was now peering out open-mouthed from a safe position behind a pomegranate tree. All that could be seen of his grandson was a tuft of black hair and two scared eyes staring from under the protection of the old man's jacket.

Señora Ferrer frowned at Ellie in utter puzzlement and crossed herself – just in case.

Ellie's normally placid disposition was now overcome by sheer frustration. She grabbed the quaking Francisca by the hand and dragged her over to the house. Gesturing furiously at the kitchen window, she blew another thunderous raspberry and yelled at our cowering neighbour, 'In the *cocina* . . . Robin and Marian, every night . . . *comprende*?'

Somehow, the penny finally dropped for Señora Ferrer. She wiped a tear from her eye and raised her shoulders almost to ear level, her elbows firmly by her sides and the palms of her hands extended forward in that uniquely Spanish 'so what's the big deal?' pose. *'Normal. Perros serán perros. Es normal.'*

She handed Ellie the parcel of dog food, tossed her head defiantly, then turned and strode off over the field towards her *casita*.

'Dogs will be dogs? Is that what she said? Well, her dogs can be *normal* on her kitchen floor, NOT on mine! I'm going to have my shower now!' Ellie slammed the front door behind her. I'd never seen her in such a rage.

When the two ladies had departed the scene, Rafael emerged from behind the pomegranate tree and shuffled cautiously up to the house, carrying two bags of oranges in one hand while shielding his grandson behind him with the other.

'Un poquito enferma, su esposa?' he asked, nodding towards the house.

'Sick? No, my wife's not sick – just a bit . . . well . . . annoyed.'

Rafael wasn't convinced. *'No está enferma?'* he enquired incredulously, screwing a forefinger into his temple.

I decided that a swift change of subject would be prudent, but Rafael had the same idea and he beat me to it. He jabbed me in the chest with a stubby, callused finger that hummed of oranges and goats. A quick-fire salvo of Andalusian dialect rattled out at me, and I didn't understand a single word of it.

'Mas despacio, por favor,' I pleaded. (Asking someone to speak more slowly was one of my most fluent Spanish phrases.)

'Ma' de'pathio? U'ted no puede entender mi e'pañol?'

Sí, sí. I could understand his Spanish quite well, I lied, but I was still learning the language and I had some trouble when he spoke too quickly. Could he speak just *un poquito mas despacio, por favor*?

Rafael pushed his cap to the back of his head and breathed in deeply through his nose. Then he was off again, pouring out a verbal flood of completely unintelligible utterances, no more slowly than before, but twice as loudly. Now I knew what it felt like to be a Spanish waiter in a hotel full of British package-tourists. Rafael's arms flailed around wildly as his speech grew faster and louder. I had the distinct impression that I was being lectured on something about which the old fellow felt quite strongly.

When he had finished, he pulled the peak of his cap back down over his eyes and squinted out at me expectantly. A decidedly pregnant silence followed.

'*Qué?*' I finally asked feebly, then immediately wondered why I had said it. I sounded just like Manuel in *Fawlty Towers*. I felt a complete idiot, and I must have looked the part too, standing there totally dumbfounded. Rafael's grandson giggled and disappeared round the corner of the house.

'*Qué?*' I repeated involuntarily, quite unable to think of anything else to say. I was temporarily numbed by the realisation of just how much we had to learn about conversing with our Spanish hosts. Neither Ellie nor I had gained any points so far that morning.

Fortunately, the old man must have realised that he too had a bit to learn on the subject of communicating with foreigners, or maybe he simply felt sorry for this particular *loco extranjero*. '*Amigo,*' he said quietly. He patted me on the arm, raised a forefinger to his lips, then added in a whisper, *'U'ted tranquilo . . . U'ted tranquilo.'*

I could understand that he was telling me to be calm, to relax. The fact that he was the one who had been ranting and raving like a lunatic seemed to have been conveniently overlooked. However, an emphatically placid Rafael took me by the elbow and guided me over to the terrace at the side of the house where he indicated that we should sit down on the two rickety chairs opposite the well. This would not have been the first time that he had rested his old bones there, I guessed.

What I needed was a tractor, he explained, tapping the back of my hand to punctuate each carefully enunciated syllable. I listened intently, humbled in the knowledge that Rafael was now talking to me as if I were an infant grandchild. Come to think of it, even his youngest *nieto* probably knew more Spanish than I did anyway.

This farm had become an absolute mess, Rafael lamented. 'Look at the weeds. *Mierda!'* Weeds drew the moisture from the soil, and that was a terrible waste. The weeds had to be ploughed under now, didn't I understand?

I appreciated only too well what he was talking about. The common sense basics of agriculture were fairly universal, I supposed. But unknown to Rafael, I had been led to believe that normal autumn cultivations were to have been completed by Tomàs Ferrer before we took possession of the farm from him. I didn't know why the work hadn't been done, and Señor Ferrer had never mentioned the subject. Perhaps I had simply misunderstood the arrangement, so I'd already resolved to tackle the unsightly green carpet myself – just as soon as I had time to purchase the necessary equipment. For the moment, it was simpler to nod in silent agreement with my well-meaning adviser.

Then there were the trees, Rafael continued gravely. *'Ah, sí, los árboles . . . los árboles.'* He stroked his chin and gazed out over the orchards which, to my untrained eye, looked quite wonderful in the bright morning sunshine.

'Surely there's no great problem with the trees?' I ventured hopefully.

Rafael shook his head. 'Not a great problem, *mi amigo*, a *very* great problem. *Ah sí. Un problema MUY grande!'*

As my knowledge of fruit trees was on the lower end of negligible, I was keen to pick up information from anyone who would take the time to tell me. I'd bought a few books on the subject before leaving Britain, and I really had tried hard to delve into the depths of arboriculture, but I had to admit that I'd hardly even scraped the surface of what seemed to be an infinitely complex science.

'So what exactly is this *problema muy grande*?' I asked Rafael anxiously.

He cleared his throat, paused, cleared his throat again, started a rasping cough which developed into a guttural gurgle that sounded like some old blocked drain being cleared, then gripped my arm tightly for leverage as he launched the resultant mucous missile in a perfect arc high over the well to land – presumably on target – at the foot of a walnut tree. Rafael smacked his lips and winked a watery eye at me. 'What a beauty, no? Not many of them to the kilo, *amigo!'*

'Yes, quite, but what about this tree problem?' I urged, trying not to appear too impressed by his obvious spitting prowess.

Rafael leaned back in the creaking old chair and clasped his hands behind his head. 'Ah, if you had only seen this place when the Señora Francisca's father was in his prime,' he reminisced, closing his eyes and pushing his cap down over his face. 'Ah, *sí* . . . Señora Francisca's father . . . old Paco. What a man . . . a maestro . . . a maestro of trees. This used to be the best farm in the valley. Everybody came to look at the trees . . . so beautiful.'

'Very good, but tell me about my tree problem, Rafael. I really need to know.'

'Then there was his wine. *Ah sí* . . . old Paco's wine. Made some of it from the grapes along this very terrace – *aquí mismo.* I used to sit and drink a bottle or two with him on the warm autumn evenings . . . on these very chairs – *aquí mismo. Ah sí* . . . if you had only seen this place then. *Ah sí* . . .'

The muted sound of snoring from under his corduroy cap confirmed that Rafael had fallen asleep.

I looked at the old man snoozing peacefully. He was dressed not much better than a scarecrow, and it was unlikely that he'd had very much in the way of money or material possessions during his entire life. But he had his large family – of which he was obviously extremely proud – he had his little house, and he had his goats. More than that, he'd had the companionship of good friends in a small rural community where people, animals, land, plants and trees still had to co-exist in harmony, each depending on the other for survival and welfare; a simple life, but, in many ways, an enviable one – particularly in such a perfect climate and in such a blissful place.

Those days were all but gone, without a doubt. The advent of mass tourism, of television, of 'progress' had seen to that. Yet, although lifestyles were changing, the valley was the same as it had always been. The mountains, the pine trees, the little orchards and the old stone farmhouses were all still there. It was the world around them that had changed.

I thought again about Rafael's memories of how this farm had once been, and I promised myself to do all that I could to return it to its former state. I was privileged to have the opportunity, so it was no more than my duty to make the effort. I knew that it wouldn't be easy, but I was sure we could do it – with a lot of hard work . . . and a lot of good advice.

'Abuelo! Abuelo! Tengo hambre!' shouted Rafael's grandson, tugging at the old man's sleeve.

'Aha, Pedrito.' Rafael pushed back his cap, rubbed his eyes and yawned. 'Sí, sí – you're hungry. Time to go home now; but first I must pay the señor for the oranges.'

I didn't like to charge old Rafael anything at all. The trees were laden with oranges for which I still had to find a viable bulk market, so a few kilos for Rafael would make no difference, particularly when he had done the picking himself. But he always insisted on paying, so rather than risk hurting his pride, I would ask him for a nominal amount.

'Cuanto es?' asked Rafael, poking a forefinger and thumb into a tiny leather purse.

'Make it a hundred pesetas per bag.'

Rafael looked at me for a moment, then lifted one of the bags. 'Aren't you going to weigh them?'

Why bother weighing them when I'm almost giving them away for nothing, I thought to myself? But, to keep Rafael happy, I took the old spring balances off the wall and suspended the bag from the hook. 'Ten kilos, *mas o menos*, Rafael.'

'*Muy bien, muy bien.*'

He seemed pleased now, and duly handed me a hundred pesetas. I suspected that Rafael really was grateful to be getting a bargain, and that his insistence on having the oranges weighed was only because he wanted to know just how *big* a bargain he was getting.

He lifted the second bag. 'How about these ones?'

'OK – I'll weigh these for you too, if you want.'

Rafael appeared slightly abashed. 'Well . . . it's just that I picked these ones off the ground . . . I thought that maybe –'

'Off the ground? But they'll have worms and things in them. You can't eat those.'

'*Hombre*, the worms are good for you. The oranges will be squeezed for juice anyway, so any worms in there will be all squashed up too. A little bit of sugar and . . .' He laughed to see me shuddering at the thought. 'I tell you, it's good for you, *amigo*. I have been drinking the juice of wormy oranges all my life, and look at me. I have four sons, three daughters and fifteen –'

'OK, OK, Rafael. I believe you, and you're welcome to all the fallen oranges you want – any time.'

'*Grati*'?'

'*Sí, sí, absolutamente gratis*. No charge.'

Rafael was a happy man. 'Just like old Paco,' he beamed as he turned towards the gate.

Although I realised that his compliment was prompted, to some extent, by self-interest, I believed that his feelings were genuine enough. In his own way, he was returning my goodwill, and I thanked him for it.

'But before you go Rafael – what about my tree problems? You still haven't explained.'

'Don't ask me, *amigo*. I know nothing about fruit trees. But I have heard them say in the village that the trees here are a disgrace now. You better ask an expert. *Adió', y mucha' grathia'*.'

'Talk about making my day,' I bellowed upstairs to Ellie as I stormed back into the house. 'That old bugger wound me up for fully half an hour about the terrible state of the fruit trees on this place, then he cheerfully told me that he knew nothing about the subject and pissed off down the lane with his bloody goats . . . and my bargain oranges.'

'Calm down, dear,' said Ellie, stepping jauntily downstairs, all dressed up and ready to go out. 'Or *u'ted tranquilo*, as your old pal would say.'

'Well, why rubbish the trees and worry me stiff if he doesn't know what he's talking about? Bloody cheek. I'll charge him full price from now on.'

Ellie handed me a glass of juice freshly squeezed from some oranges which she had gathered the previous evening. 'Here – sip this. It'll cool you off. And no more swearing. We've had enough bad language *and* bad feelings around here for one day, thank you very much.'

'Well, advice is one thing – God knows, I could do with plenty of that – but slagging off the fruit trees . . .'

'No, be fair. I'm sure the old boy didn't mean to be offensive. I was listening from the bathroom window, and he was speaking so slowly most of the time that even I could get the drift of what he was saying.'

'Uh-huh?' I scratched my head and cast a quizzical glance at Ellie over the top of my glass.

'Yes, I did understand quite a bit, so don't look so amazed. Think about it. You've noticed the sticky black stuff on some of the oranges, and all the dead and overgrown branches. Even you know that a lot of trees need attention, and you're a complete novice. Right?'

'Maybe, but they don't look that bad. Let's face it – that's a pretty heavy-looking crop of oranges out there.'

Ellie looked out of the window and nodded pensively. 'Hmmm, I suppose so, but how much better could it be? That's the point. And I think Rafael was only marking your card – just stating the obvious, in case you didn't know.'

'OK, but the neighbourly thing to do would be to tell me where to get some expert advice instead of putting the wind up me and then disappearing off down the lane like that.'

'No doubt Rafael could have given you plenty of advice on the trees himself. I doubt if there are many old blokes in these parts who couldn't. But not all of them will be known as the *real* experts.'

'The '*maestros*', you mean?'

'Right. And it could be that Rafael isn't noted as a fruit tree maestro. Goats, maybe – but fruit trees, maybe not. So you should go to a real tree expert – that was all he was saying. That way, he wouldn't be stepping on anyone's toes . . . and he couldn't be accused of giving you duff advice either. Makes sense to me.'

I felt the gloom lifting.

'Ellie, you've done it again. You've restored my faith in human nature. An object lesson in common sense.' I gave her a pat on the bottom and drained my glass. 'Wow! That's the best orange juice yet. Which trees did you pick them from this time?'

Ellie's face lit up into a mischievous smile and she sauntered away towards the door.

'Trees?' she called back over her shoulder. 'Trees? I didn't get them off the trees. Oh no, I picked them off the ground.'

* * * * * * * * *

Like many inland towns situated near the periphery of the island, Andratx has a sister town – the Port, *el Puerto* – on the coast about a couple of miles away; a feature of Mallorca dating back to pirate times when marauding ships from the Barbary Coast of North Africa prowled this stretch of the Mediterranean. To warn the islanders of the approach of pirates, fires would be lit on the stone towers, the *torres*, which still stand watch on prominent clifftop positions round the coastline. The local fishermen and farmers would then retreat to the more easily defended inland villages where their houses were clustered round the fortified church – invariably built on the higher ground.

When we first moved into Ca's Mayoral, the narrow road from Andratx to Puerto Andratx still followed the twisting course of *el Torrente*, the torrent in name, but really little more than a dried-up ditch, except for the occasional flash flood following an unusually severe thunder storm. Leaving behind the lush orange groves and flower-draped walls and patios of the old houses on the edge of Andratx town, the valley meanders and widens into a miniature plain, protected on either side by the pine-green arms of the mountains, their lower slopes peppered with little stone farmhouses the colour of weathered straw which sit amid tiny fields of vegetables and cereals, and ancient, rubble-walled almond plantations which chequer the landscape in random enclosures of gentle, unembellished beauty.

While the risk of encountering a band of pillaging pirates no longer exists, every bend in the tortuous lane may still conceal the threat of mortal danger for the modern traveller. Large tanker trucks run a constant shuttle service between nearby fresh water wells and the many households in the area that have no mains supply or private well, and speed is the byword of the brawny truckers as they wrestle their lumbering rigs around the tight bends at optimum velocity.

So, the law of averages should dictate that, sooner or later, a water tanker, a car, a bus, a tractor or a donkey and cart, even a flock of sheep or one of the troops of geriatric German hikers that patrol that road in winter would contribute – in some combination – to the inevitability of a horrendous accident on one of those blind bends. But, miraculously, it hadn't happened yet, and one could only hope that the miracle would continue.

Yet evidence of one apparent near miss was there for all to see on that December Sunday. As we rounded one particularly sharp corner, there in front of us, some ten yards into a field and a good eight feet off the ground, was a little Seat 600 car, hanging by its rear wheels from the bough of an almond tree. How it had arrived in reverse and intact halfway up a tree was a mystery. There were no skid marks on the road and no tyre tracks in the field. We could only speculate that the intrepid pilot had been on a high-speed burn-up along the road on the previous night and had, perhaps, encountered the awesome bulk of a two-thousand-gallon water truck bearing down on him head-on as they both claimed the same racing line round the corner. A prudent evasive swerve onto the raised verge may well have propelled the little car over the wall and into a near-perfect trajectory for a safe landing (albeit head-over-heels) in the almond tree.

To us, that seemed the most likely reason for the phenomenon, but other more imaginative theories explaining the mystery of *el coche en el almendro* were rife among the customers in Margarita's busy newspaper shop in Puerto Andratx that morning.

One bespectacled lad reckoned that flying saucers must have been involved. A squeaky-voiced old woman instantly dismissed that idea as being much too *fantástica*, and confidently claimed that the car had fallen from a tourist aeroplane. The island had never been the same since all those jets started to fly in, she reminded her nodding audience of fellow black-clad matrons. *Dios mío*, even the weather had changed for the worse with all those *máquinas del diablo* making holes in the clouds! The group of elderly ladies crossed themselves in concert, and the motion appeared to be carried without need for further debate.

I felt an elbow in the ribs.

'Bullshit! All bloody bastard bullshit, oh yes,' came the confidential information in fairly good English bad language, spoken in a curious Hispano-Birmingham accent with faint echoes of Karachi by a slightly-built Mallorcan man in working clothes who was standing by my side in the queue. 'That bloody car be belonging to one bunch of stinking hippies, I tell you, man. They living in one bloody pigsties *casita* up

S'Arraco valley. Bloody S'Arraco valley being bloody full of them buggers. Is ridickliss! Oh yes.'

'Oh yes? Well, that's intere–'

'And I telling you this, man,' he chuckled softly, moving in closer to reveal the exclusive stuff, 'they been coming down here last night for picking up the bloody bastard Bob Hope off a boat.'

'Bob Hope? Here?'

He nodded smugly. 'I am knowing everything around here.'

'But why would a bunch of hippies be meeting a film star off a boat in Puerto Andratx?'

He closed his eyes and shook his head in exasperation. 'No bloody bastard Bob Hope – Bob Hope! I be talking to you about bloody bastard Bob Hope – dope, hash, shit, tea, space grass, can o' piss! Bloody 'ell! What planet you been living up?'

'Right – I get it now. Bob Hope – dope. The hippies were picking up some cannabis off a boat, right?'

'Is right! That's what I be saying . . . can o' piss! And they been smoking it – much joints, I tell you – and drinking the rotten guts brandy behind the fish market all bastard night. I been being watch them from the Bar Acal, oh yes, and they been being so bloody high on Bob Hope and Mahatma Gandhi when they been leaving in that car that they been floating up that bloody bastard tree! Is bloody ridickliss!'

My informant laughed aloud and stepped up to the counter to pay the impassive Margarita for his paper.

'You English?' he called back to me on his way out.

'Ehm, no – Scottish,' I replied, feeling suddenly exposed and conspicuous as the assembled array of local faces turned towards me.

'All the bloody same,' came the confident response from the doorway. 'English, Scottish, Irish, Welsh – all the bloody same to Jordi. I been being there.' He laughed again and disappeared between a cluster of magazine stands outside the shop.

Ellie was waiting for me at a table outside the tiny Bar Tur, just along the way on the corner of the little elevated plaza which overlooks the wide, picturesque sweep of the harbour. It had become an essential part of our Sunday mornings to sit there for a while, reading the *Majorca Daily Bulletin*, having a leisurely drink in the soothing warmth of the winter sunshine, and never tiring of gazing through the palms at the spectacular views over that magnificent, sheltered bay.

The port's little fleet of sea-going fishing boats was lying alongside the quay just below us, their distinctive flared bows and the sweeping curves of their hulls painted the brightest shades of blue, red, turquoise or green, their decks bristling with a forest of beflagged marker poles that sprouted from orange-coloured, globular floats like festoons of balloons and bunting hung out for a fisherman's fiesta. Tied up nearby were some *llauds*, the little Mallorcan inshore fishing boats with such beautifully flowing lines of age-old Mediterranean style, all pristine white – in contrast to the vivacious party attire of their big sisters along the quayside – and reverentially bearing the names of their owners' wives, sweethearts, canonised Carmens, Catalinas and Marias, or the guiding stars of the sea. Beyond these working boats, the tall masts of countless anchored yachts swayed in a silent water ballet on the gentle swell, and far over the glistening water, on the other side of the port, the encircling mountains rose steeply from the shore, their green woods dotted with white-walled villas clinging to the sheer slopes like clusters of petrified Alpine flowers glinting under a Calabrian sky.

This was Puerto Andratx at its best – quiet and sleepy, and recalling the tranquil ambience of its bygone days as a simple, undiscovered fishing hamlet, nestling peacefully on the safe shores of this most favoured of havens. But this idyll was the very antithesis of what the Puerto would become in the months of high summer. Then, rubber motor-dinghies overflowing with smooth-handed city sailors from those same winter-abandoned yachts would decant ashore in guffawing, hoop-shirted gangs and submerge the quayside tables of the little bars and cafés in nightly cacophonies of Anglo and Saxon heehaw tales of the day's great adventures on the champagne-spuming main. *Coño!* – as more than one hoary local fisherman had been heard to mutter on finding their favourite dominoes table at their favourite bar occupied night after August night by the summer mainbrace-splicers – their Andratx ancestors had been lucky that they only had the Barbary pirates to contend with! *Prou!*

Winter invasions were likely to be no more bothersome, however, than a coachload or two of Palma pensioners stopping off for a pre-lunch *paseo* along the waterfront – the elderly ladies in intricately-crocheted shawls and immaculate hair-dos strolling arm-in-arm in seemingly deep and spirited conversation, while their *maridos* followed a respectable distance behind, resplendent in Sunday-best blue suits and polished brown shoes, or even, in the case of some groovier old guys, in a pair of multi-coloured sneakers to cut a generation-bridging dash beneath those carefully pressed serge trousers. The signs of change were everywhere.

After Ellie had scoffed her mandatory Sabbath *ensaimada*, a light Mallorcan pastry, we followed just such a party of old Palmesanos along the quay, pausing now and then to watch the fishermen unhurriedly preparing their boats and tackle for the next day's trip, while singing to themselves or shouting jokes and banter from boat to boat in their slow *mallorquín* drawl, and recreating a daily scene unchanged by centuries.

With no warning, as can happen in the Mediterranean at that time of year, a fresh breeze began to blow in from the sea, hurling foaming breakers against the harbour wall and compelling the strolling Palma matrons to turn their backs to the offending wind while whipping their protecting shawls over those precious coiffures. Only a few weeks earlier, we would have been delighted to stand there and enjoy the spectacle of the waves urging the moored boats into a livelier dance, but now, we were pulling on our sweaters and making for the shelter of the car – almost like seasoned natives. Yes, we would lunch inland today.

Tucked away in its own grounds in the forested foothills of the Sierra Burguesa just west of Palma is Son Berga, an old, fortified farmstead. Its thick walls of mellow stone are surmounted by stout castellations, and roses and jasmine now climb past defensive arrow slits that are, like the coastal watchtowers, another reminder of the island's turbulent history. Pink clematis and purple bougainvillaea now scale the ramparts unopposed and clamber over the moss-brindled, ochre undulations of the ancient tiled roofs. Today, the Restaurante Son Berga occupies the main building of the farm, a cloak of ivy covering the wall by the geranium-covered flagstone steps leading up to the entrance.

'You are just in time, *señores*,' said the waiter, showing us to a table in a quiet corner of the barn-like dining hall. 'We have a wedding party today, so we are going to be even more busy than usual.'

The customary basket of crusty, brown bread and a dish of farm-pickled olives (complete with stems and leaves) was placed before us, with an additional offering of strips of raw carrot with a tangy cheese dip and complimentary glasses of muscatel wine to sip while we studied the bill of fare and savoured our surroundings.

Son Berga had all the attributes of the definitive image of a rustic Mallorcan eating place – sturdy sandstone pillars supporting a vaulted ceiling on heavy, exposed timbers; tawny stone walls stained with the warm patina of smoke drifting from great logs smouldering on the open fire; shafts of sunlight, filtered and softened by the smoke, slanting through the small, deep-set windows and etching fan-shaped parquetry patterns of shadow through an assembled army of chunky table and chair legs onto the aged floor tiles. It felt cosy and welcoming, like an olde worlde Christmas card tavern, and, understandably, its reputation for serving traditional Mallorcan food of the highest quality had made it a popular weekend haunt of many upwardly-mobile Palma families.

The casually-but-expensively-attired business and professional types who were starting to arrive with their sartorially-matched wives and children could feel 'back home' here, at one with their roots – as could the once-rural grandparents who tagged along gladly (if a little self-consciously) to complete the line-up of the celebrated Spanish extended family. They were here for their weekly fix of old Mallorca on a plate, beautifully prepared and cheerfully served by attentive waiters in crisply-starched, long, white aprons. For them, Son Berga was the best of all worlds – a place where the socially-ascending generation could observe and be observed by others of like genre, where their offspring could behave the way that offspring do, and where the old folks could take an edible freebie trip down memory lane. Ellie and I just liked the food and the atmosphere.

'*Para comer, señores?*' enquired our waiter, flicking open his note pad.

I asked him for *Arroz Brut*, a house speciality which we had both agreed to start with, then waited for Ellie to order her own main course in her best restaurant Spanglish, the words of which she was now inclined to weigh carefully following a recent shopping incident in Andratx.

Her feminist determination to pedantically refer to a chicken as a female *pollA*, as opposed to the paradoxically-accepted usage of the masculine *pollO,* had never drawn more than a few suppressed titters from the village ladies standing in line behind her in the local butcher's shop. It was only after successive occasions of asking the butcher for *una pollA grande* that a young, babe-in-arms mother came forward and whispered coyly in Ellie's ear that to use the feminine *pollA* in this context was *incorrecto*. *La señora* was making *un error grave*. *Un pollO* was a table chicken, irrespective of sex, she explained, and *una pollA* was – ahem

– how could she put it? *Sí, una pollA* was – She pulled down the front of her baby's nappy, flicked out the little fellow's willie and waggled it with her forefinger. Smiling proudly, she declared, '*Hombre!* THIS is *una pollA!*'

The diminutive butcher then made a swift and unscheduled visit to his back shop – perhaps to spare Ellie's blushes, or more likely to spare his own. Being a long-time plyer of the meat trade, the butcher knew full well that his housewife clients also appreciated the merits of good hanging . . . and that baby boy *was* precociously and enviably well hung!

'*Carám!* Hung like a donkey,' one aged crone cackled on hobbling up to have a closer gander at the object of the impromptu language lesson. *José, Maria y Jesús!* She had never seen no *bambino* with such a *pistolita*. *Nunca!* Fortunate would be the *chiquita* to have her *fuego* attended to by *his* fire engine in a few years time!

Right on cue, the *bambino* kicked his feet in the air, googy-googied gleefully and squirted a glorious, pulsating fountain ceilingwards before his mother could readjust his nappy.

'HOO-EE-EE!' screamed the side-stepping line of lady shoppers. '*Viva el bomberito!* Long live the little fireman!'

From that day on, Ellie always made a particular point of accentuating the 'o' ending in *pollo* when ordering chicken in a restaurant. And who could blame her? To have repeated her *error grave* of the butcher's shop could have proved decidedly tricky if any waiter had chosen to take her order literally. What a culinary cock-up that would have been!

She was risking no such piddling linguistic clangers at Son Berga, however – prudently steering well clear of chicken dishes and confidently ordering *Gazapo* – 'With *mucho* bits of *tomate* and, ehm – onions on the side, but no *mucho* bits of bread – *pan*, that is, *por favor*.'

Shrugging slowly, the waiter scribbled furiously.

'You're quite sure you know what you've ordered?' I asked out of the corner of my mouth, not wishing to dent Ellie's pride in her expanding knowledge of Mallorcan menuspeak.

'Of course I do. Just because it's winter doesn't mean that I can't choose something summery and salady, does it?'

'Fair enough – as long as you know what you're doing. I think I'll stick with the When in Rome principle and go for the *Escalopes mallorquines*.'

'*Para beber, señor?*' asked the waiter, gesturing now towards the wine list.

'For a Sunday treat, I think I'll have some of Mallorca's best – a bottle of José Ferrer Gran Reserva. Oh, and a bottle of still mineral water, *agua sin gas*. *Gracias*.'

A continuous row of tables, occupying one complete end of the restaurant, had been prepared for the wedding party, and at strategic points – all within an arm's length of any seat – clusters of bottles and jugs had been assembled. There was wine of the three complexions, water by the litre, and the essential *gasiosa* – the local version of fizzy lemonade, much favoured as a diluter (or, more truthfully, softener) of the more 'robust' red wines dispensed from small barrels on the gantries of most Spanish bars. Between two of the arrow slits on the wall, a large mural, depicting a group of traditionally-dressed Mallorcan girls posing primly with water amphoras in a pretty, cypress-sheltered garden, looked down on what appeared to be the perfect setting for a joyful wedding feast.

Then the guests started to file in – and what a miserable-looking bunch they were. If the waiter hadn't told us otherwise, we would have assumed that they had just come from a funeral. No one spoke. They just took their places and sat there in glum silence. One woman, whom we took to be the bride's mother, had been crying so much that her face resembled a burst tomato – and something told us that she had not been crying tears of joy, either.

Although the principals were nowhere to be seen, the light of flash bulbs flickering through the windows suggested that the happy couple were having some wedding day photographs taken in the little patio behind the restaurant.

'I can't wait to see the bride,' Ellie fluttered, bubbling with enthusiasm. 'Oh, I bet her dress will be beautiful – all white satin and old Spanish lace and things.'

'It's a pity you can't transmit some of your excitement to the families and friends over there. I thought Mallorcan weddings were supposed to be real rip-roaring affairs, but I've seen people having more fun in the waiting room at the dentist's.'

'Look! They're coming in now. Oh, it's so romantic! Look – isn't she beaut . . . oh, my God!'

There, doing her best to hide behind her doomy groom as he slouched dejectedly to the table, was the reason for the mass despondency – the bride, looking distinctly unradiant in her delicate blue maternity dress which did nothing to conceal the swollen evidence of a deadly trouser-snake strike. Her mother began to bubble again.

'I hope she stood behind a tree when the photographs were being taken,' I muttered. 'She must be seven months gone if she's a day.'

'Don't be so insensitive! It's not a crime to be pregnant, you know. And her boyfriend – her husband – is just as much to blame, if not more. I mean to say, she could hardly get in that condition by herself, could she!'

'Well, there's always immaculate conception,' I said, glancing up at the crucifix above the door. 'Maybe the poor guy was framed.'

Ellie's silence said everything.

Over at the wedding reception, the ice was broken at last by the sound of the bride's mother whimpering. Her husband, clearly sick to the teeth of her continual sniffling, grabbed a bottle of wine and poured himself a tumblerful. Throwing it down his throat in one gulp, he stood up, tumbler in one hand, bottle in the other, and proposed a half-hearted toast to the disconsolate couple. '*A la novia, y al novio.* To the bride and groom.'

The guests raised their glasses. The father of the bride raised his bottle and, casting a fleeting glance at his daughter's abdominal protuberance, mumbled, 'And to the *bambino!*' He slumped back into his chair. His wife wailed with added gusto.

By this time, the restaurant had filled to capacity and, following the shocked silence which had greeted the bulging bride, a cheery clamour was developing throughout the room as each family group settled into the noisy ritual of Sunday lunch Mallorcan-style. Olives and bread were being stuffed into mouths, the wine was flowing freely, and the cross-table conversations grew even louder and more animated.

A great time was being had by all – all except the wedding party, that is. A black cloud had settled over the condemned couple, and its shadow was cast over the entire company of guests.

We had identified the groom's mother – a skinny creature with greying hair drawn severely into a bun, accentuating the bird-like appearance of her sharp, bony features. She sat motionless, her hands firmly clasped in her lap, her narrow, black eyes firing arrows of loathing over the table at her son's new mother-in-law, who was now attempting to drown her sorrows by chain-drinking glasses of *vino rosado*. Her husband was doing his best to encourage her, being well into the second bottle of his preferred *vino tinto* himself.

Soon the steaming *greixoneras*, those ubiquitous and versatile earthenware casserole dishes of Mallorca, started to appear from the kitchen, balanced aloft by sure-footed waiters nimbly and good-naturedly dodging the toddlers playing between the crowded rows of tables, the great vessels brimful of piping-hot *Arroz Brut* – clearly a popular dish at Son Berga. The name *Arroz Brut*, or 'Dirty Rice', is a rather unflattering description of a classic Mallorcan peasant broth of rice in a creamy, saffron-golden stock with meat, vegetables and the magic ingredient: *los caracoles*, those tiny, tender Mallorcan snails in their dainty, tabby-coloured shells.

On the basis of saving the best till last, we left the *caracoles* in our *greixonera* until we had finished every last drop of the delicious *Brut* – a fabulously filling task in itself, exacerbated for me now by Ellie's sudden aversion to snails, one of which, she claimed, had retreated horns-and-all inside its shell when she threatened it with her spoon. I was obliged – under the watchful eye of our waiter – to show pleasure in profiting from what was (I fervently hoped) the product of her over-fertile imagination.

Following the example of the surrounding experts, I winkled out succulent snail number one with a toothpick and dipped him into the accompanying saucer of *All-i-oli*, a spoon-supporting sauce of raw egg yolks, olive oil and garlic, and garlic, and garlic.

'*Más vino* – you must drink more wine,' announced the waiter, flamboyantly replenishing my glass with the José Ferrer. 'You must drink *mucho, mucho vino* with the snails, *señor*. It is the Mallorcan custom. The *All-i-oli* purifies the souls of the *caracoles* and the *vino* sends them to heaven happy. *Más vino!*'

Whatever the *All-i-oli* did for the snails' souls, only a snail could say, but it certainly made my eyes water and it wasn't hard to understand why the drinking of *mucho, mucho vino* was tied in with the eating of *caracoles* and *All-i-oli* – it was the only way to prevent your tongue from blistering.

'*Adiós, caracoles,*' I wheezed, washing down another nippy mouthful of garlic-embalmed gastropods. 'Have a pure and happy trip, lads. *Vaya con Dios!*'

'*Bravo!*' shouted the waiter, thumping my back. That was the way to eat snails, he enthused. 'Go for it!'

Buoyed up by his adept incitements and inspired by the deluding dizziness of the wine, I avidly dredged the *greixonera* with my spoon until every last snail had been scooped out, plucked from his shell and dispatched down my gullet.

'Ahhh, that was good,' I burped, breathless and moist-eyed, to the practised admiration of the waiter. '*Fabuloso!*'

'You're disgusting,' said Ellie. 'How you can bring yourself to gobble up those poor little creatures like that, I really don't know.'

Her expression of revulsion was nothing compared to the look of consternation which crossed her face when the waiter presented our *platos segundos*.

'This is not *Gazpacho*!' she gasped, her frown flitting between the waiter's confused face and her plate, where lay a pelt-shaped piece of meat that had the appearance of a sky-diving, skinned kitten, the tender flesh glistening in a terracotta glow to match the plate, the sear marks of the grill standing out like charred tiger stripes across the tiny, semi-cremated carcass. 'I didn't order this! I ordered *Gazpacho* – chilled tomato soup . . . with bits of tomato and onion and things to sprinkle on the top!'

The waiter raised his eyebrows and pulled his mouth into a lop-sided smile as the veil of confusion lifted. *'Señora,'* he crooned, placing a menu in front of Ellie and patiently running his finger down the list. '*Gazpacho* we do not have, *no*? But *Gazapo* we do have, *sí!*'

Ellie toyed with her chin and silently mouthed the words, *'Gazpacho, Gazapo, Gazpacho, Gazapo* . . .' her look of consternation now bordering on total mystification. 'So, what's the difference?'

The waiter pointed to the relevant word on the menu, then at her plate. 'This is *Gazapo, no Gazpacho. Comprende?'*

Ellie looked blankly into space. *'No comprendo.'*

Taking a deep breath, the waiter massaged his hands like a concert pianist preparing for a performance. 'OK, madame – I gonna esplain you sonseeng een Eenglees, OK? Thees deesh ees *Gazapo*, and thees you order for sure, because I already esplain you – no ees *Gazpacho* on the menu.' He glanced over at me with a palm-upward gesture of one hand. 'OK, *señor?*'

'*Correcto,*' I confirmed.

'OK. So now I esplain you proper, madame. Thees ees the *Gazapo* what you order, OK? But the *Gazapo* no ees the ice-cold tomato soups. That ees the *Gazpacho* what we don't got. What we do got and what you got ees the *Gazapo* – the red-hot baby rabbit!'

Ellie puckered her lips and nodded resignedly: 'Now I *comprendo.*'

Just how she could bring herself to gobble up *that* poor little creature when she had baulked at the *caracoles*, I really didn't know.

As for my *Escalopes mallorquines*, all I can say is that they were a tribute to the universal reputation of nonpareil excellence enjoyed by Mallorcan pigs for centuries. Both dishes were served with vegetables – and *mucho* bits of *tomate* and onion on the side for Ellie.

For me, this was Mallorcan country cuisine at its unfussy, wholesome best. Ellie tactfully said nothing, but it was quite obvious that she had thoroughly enjoyed eating the result of her latest linguistic gaffe. And why not, indeed? Judging by the number of kids in Son Berga who were also tucking into *Gazapo*, it was plain to see that we were a long way from Beatrix Potter Land, anyway.

Just then, there was a loud peal of laughter from some children at the next table where their tiny, silver-haired grandmother had been trying for some time to coax a stubborn infant in a high chair to eat little morsels of food from a plastic teaspoon. She had employed all the usual persuasive tactics – the train puffing into the tunnel, the aeroplane flying into the hangar, even the humiliating 'Ooh-nimmy-nimmy-nimm' routine – but all to no avail. The little fellow was not impressed. He merely deepened his scowl and kept his mouth tightly shut.

Then, to the accompaniment of the older children's helpless laughter, the little old lady's selfless patience had finally been rewarded by her baby grandson giving her a skull-denting whack on the forehead with a heavy soup spoon which had been handed to him by his mother – presumably as a learning aid. Even if he was a bit slow to grasp the orthodox usage of a spoon, that infant was already showing that he certainly was a quick learner when it came to the vital business of getting his own way in life – a talent that he had surely not inherited from his granny. His mother looked proud. His granny looked semi-conscious. The rest of the family momentarily interrupted their *paella* wolfing to have a quick giggle, then reverted to the eccentric Mallorcan Epicurean practice of tearing the heads off prawns and voraciously sucking out whatever minute organ may lurk in that lowly crustacean's inconsiderable cerebral cavity.

'So much for the happy-families concept of the Spanish Sunday lunch,' I quipped to Ellie. 'That old dear would have been safer going on the coach trip to Andratx with the rest of the pensioners.'

'Rubbish! It's nice to see all the generations eating out together. It's a great tradition, and I envy them.'

'It's good to watch – I'll give you that.'

At the other end of the hall, meanwhile, little was being done to uphold the great tradition in the new, soon-to-be-extended family. The grieving groom had been pushed to his feet, and we presumed that this would herald the onset of some painful speech-making; but mercifully not, as it happened. The local custom, we learned, was for the groom to publicly kiss his bride for the first time at this stage in the proceedings, when sufficient wine should normally have been consumed by the guests to ensure that the ritual kiss would signal the start of the real jollifications. Accordingly, the bloated bride was hauled to her feet by two sturdy maids of honour jacking her up at the elbows. Then, to some compulsory but spiritless cutlery-banging and cat-calling, her reluctant spouse planted the matrimonial kiss on her rigidly closed lips.

The groom's mother looked as if she was about to throw up. The bride's mother looked as if she already had. Probably unaccustomed to the considerable quantities of wine which she had guzzled since her daughter's ignominious entrance, the wretched woman had gradually slid lower and lower into her chair, and was now sunk into a loosely-paralysed position with only her wobbly head visible above the table. Her dainty wedding hat of delicate pink flowers had settled at a jaunty angle completely covering her right ear, the pastel shades of the petals curiously complimenting the pale green of her face. Her husband was beyond caring.

A gloomy silence descended upon the company once again.

'You know, Ellie, with all the booze and bad blood flowing over there, you'd have thought that at least one good-going punch-up would have been on by now – but nothing.'

'Says a lot for the placid nature of the Mallorcans.'

'Nah – it's not normal. It's a farce. Why spend money on all this reception sham? Better to let the silly young buggers go home and be miserable in private.'

'That really takes the biscuit! Haven't you got any romance in your soul?' Ellie exclaimed, fiddling ominously with her glass of water.

'Hey, look!' I shouted, nodding towards a timely distraction at the wedding table. 'Somebody just cut the bloody bridegroom's tie off!'

'Oh, no – that's not funny. As if he hasn't got enough to cope with.' Ellie was visibly distraught.

'Do not worry, *señora*. It is the Mallorcan custom,' said our waiter, pausing en route to the kitchen with an armful of dishes. 'See how they cut the tie up and pass the pieces around? All the guests pay a little for each piece of tie, and the money goes towards the honeymoon. That is the custom *normalmente*, but on this occasion . . . well, I do not think the guests will cough up one peseta.'

'But why not, for goodness sake?' asked Ellie.

'Because, *señora*, everyone can see that this couple have had their honeymoon already!' He laughed heartily and clattered off through the kitchen doorway, only to reappear a few moments later with a cargo of full plates. 'Do not look so sad, *señora*. It could have been worse. The Mallorcan custom also calls for the garter of the bride to be removed and cut up, but on this occasion . . .' He chortled and looked over at me. 'On this occasion, *señor*, that bride will be unable to bend down far enough to get the garter off by herself, and no one is going to help her, because . . .' Sensing that Ellie might be offended by his observations, he flexed his knees until his mouth was level with my ear and whispered,

'*Hombre*, because it will be a while before anyone is interested in sticking his hand up her skirt again – garter or no garter, no?' He flashed me a dirty wink and continued on his busy way.

I tried not to let Ellie see me smirking.

'It's all right. Snigger away,' she huffed. 'Be my guest . . . and don't bother to tell me what the waiter said. I'm just not interested!' She turned her head away contemptuously, only to be confronted face-to-face by our returning waiter.

'*Por favor, señora*, I hope you were not displeased by my little joke about the bride.'

Ellie forced a polite smile. 'No, no – that's quite all right. Ehm, in any case, I didn't quite hear the punch li–'

'You see, *señora*, the bridegroom is my brother.'

'Oh dear, I am truly sorry . . . I mean, congratulations . . . well, what I really mean is . . .' Ellie was in a flap.

The waiter smiled sympathetically. 'It is my *madre* who has created the bad atmosphere,' he confided, looking towards the gaunt figure of his stern-faced mother. 'She has had a hard life bringing up five of us alone since my father died when we were small, and now my young brother . . .' He shrugged. 'But no *problemas* – she will accept the bride and her family in time, and when the *bambino* arrives . . . she will be like all doting grandmothers. No *problemas*, eh! Now I will go to fetch your *cafés*.'

I glanced over at the little old lady at the next table, who now had a bright blue, spoon-shaped welt on her forehead for *her* doting grandmotherly troubles.

'There you are, Ellie,' I said, gesturing in turn to the spoon-battered granny and the shotgun wedding party. 'What have you got to say about the great tradition of the extended Spanish families now?'

'All I said was that it's nice to see them all eating out together. And I'm sure that old granny would rather be here – even with a lump on her head – than sitting at home alone, watching the telly like so many families leave their oldies to do in our country. We should take a leaf out of the Spanish book.'

'Right, so the first time your mother comes over to visit, we'll bring her here for Sunday lunch, OK?'

Ellie's eyes narrowed, and she fingered her water glass again. 'What are you getting at?'

'Nothing – it's just that I thought it would be a good chance to take a leaf out of that Spanish nipper's book and knock six bells out of her head with a ladle. We could tell her it's the Mallorcan custom.'

'OK – that's the last straw! First you didn't tell me that I had ordered Barbecued Bright Eyes as a main course –'

'But I didn't know that *Gazapo* meant baby rabbit – honest!'

'Yes you did! I can tell when you're having a ball to yourself at my expense. Then you took the mickey out of those poor newly-weds, jeered at the old granny's lump, and now you've threatened to murder my mother. You've asked for it!'

I started to laugh. I didn't want to, but I couldn't help myself. 'No, Ellie, don't . . . don't throw the water,' I pleaded, holding up my hands. 'It'll get into my wine and – '

'Never mind the wine. It's you that I'm after, and I know just how to wipe the smile off your face in public, don't I!'

There was a wicked little glint in her eyes as she lunged over and grabbed me by the ears, pulling me towards her, forcing me to stand up and stretch across the table where she fixed an inescapable smacker on my protesting mouth.

As endless moments passed, I became aware of a growing silence settling all around us. I opened one eye and squinted round the restaurant. Dammit! All eyes were on us. I tried to pull away, but Ellie maintained her vice-like grip on my ears. Then it started – quietly at first, but growing louder and louder until the entire hall was reverberating to the steady beat of a Spanish slow handclap. The noise built up as feet were stomped and tables thumped in rhythm, the volume rising to a climax and culminating in an ear-splitting cheer as Ellie finally released me from her marathon kiss.

I dropped back into my chair, my head reeling. Ellie looked coyly down at her lap as the applause continued, an unmistakable smile of quiet self-satisfaction in her expression. I was beside myself with embarrassment, fumbling nervously at my tie while half rising from my seat in awkward acknowledgement of the howls of acclaim from the other tables: *'VIVA EL AMOR! VIVA LA PASIÓN! VIVA-A-A!'*

'Oy! Bravo, señor!' yelled our waiter, thumping me on the back and grinning from ear to ear. 'They say you *británicos* are not as hot-blooded as we Latins, but what can I say? You really showed my young brother how to do it. And look – you have even made my *madre* happy!'

I risked a brief glance over at his mother, and bugger me if she didn't have a twinkle in her eye, inclining her head towards me like some coquettish sparrow hawk, and raising her glass of *gasiosa* in salute. I looked away rapidly. Hell, that woman must have been a widow a long, long time!

'Hey! A *caballero* with much power over the *señoras*, your husband, no?' the waiter said, winking at Ellie and making a guardedly suggestive upwards movement of his fist. 'It is the bewitching effect of the snails and the garlic. They make a man very *atractivo* – give him *mucha atracción sexual, sí?*'

'To another snail, maybe,' Ellie mumbled, fanning her nose with her hand.

Still smiling, the waiter placed our coffees on the table, together with two small glasses and a couple of bottles of green liquor which appeared to have bonsai fir trees growing inside.

'Please be my guests in a *copita* of our famous Mallorcan liqueur, the *Hierbas*. It takes its flavour from the wild mountain plants in the bottle – the camomile, the rosemary, the fennel. One bottle contains the sweet *Hierbas*, the other the dry; but for me, the best way to drink it is fifty-fifty – *meech-y-meech*, as we say in *mallorquín*.' Then, winking again at Ellie, he whispered, 'The *Hierbas* has many magical properties, *señora* – an elixir of life, some say; a mosquito repellent, for sure; but most *importante* . . . *un afrodisiáco potente, un afrodisiáco irresistable, eh!*'

'A pity for his bride that your young brother hadn't signed the no *caracoles* and *Hierbas* pledge about seven months ago, then,' she said under her breath.

'You have no family, *señores?*' the waiter asked, carefully measuring equal amounts of *Hierbas* from both bottles into our glasses.

'*Sí*, two sons,' Ellie replied. '*Dos hijos.*'

'Only *dos hijos!* Oh, *señora!* And with a husband so *amoroso* as yours? Hey, better to have a *familia grande* like all of us *mallorquines* – many sons to keep you when you are old, no? And I think maybe this will happen – now that the *señor* has been introduced to the *caracoles* with *all-i-oli* and the *Hierbas*. *Sí, sí – es inevitable!*'

'Don't forget the rabbit,' said Ellie dryly.

'*Absolutamente!*' guffawed the waiter. 'You are thinking like a *mallorquina* already! *Sí*, eating the *gazapo* will have made the *señora* very . . . ehm, very *productiva*, very *fértil*, no? Oh-hohhh, *amigo*,' he growled, elbowing me on the shoulder, 'lucky for you that your wife made the mistake about ordering the *gazpacho*. Nobody never heard about no *muchacha* performing like no plate of cold soup. But a rabbit? *Oiga!* Different story, *señor*.'

'Well, cheers!' I called as a rapid digression, noticing Ellie's eyes narrowing again, and fearing for the well-meaning waiter's safety.

'*Salud, señores*, and *muchos bambinos, eh!* I leave for you the *Hierbas* so that you can help yourselves.'

'You're too kind,' said Ellie blandly.

'No, no, *señora*. It is the Mallorcan custom. *Salud!*'

'*Salud!*' chorused the *paella* and the battered granny family at the next table. '*MUCHOS BAMBINOS!*' called the groom's mother.

'Thank God he's left the hooch,' I puffed, sampling the herby, anise-smelling concoction. 'I could do with plenty nerve-settling juice after that ordeal. What the hell took you to play a silly prank like that? Jesus Christ – I've never been so affronted! Look – I'm still shaking like a bloody leaf!'

Ellie was in stitches. She couldn't even speak.

I quaffed another two *copitas* of *Hierbas* while Ellie pulled herself together, then I suggested quietly that it might be wise to leave discreetly, now that all the other customers appeared to be fully absorbed once more in their raucous after-lunch chin-wagging.

In hushed tones, I asked the waiter for the bill, paid him with as little ceremony as possible, and thanked him quietly for his hospitality and generosity.

Instantly sizing up my motives, he latched onto my furtive manner and whispered back, 'The pleasure is mine, *señor. De nada.*'

I shook his hand unobtrusively and ushered Ellie stealthily towards the exit. I was tip-toeing as we reached the door, opening it gingerly and not even daring to look back as I stepped smartly aside and shooed her through.

'*OLÉ!*' came the concerted roar from behind me, and I spun round to see our supposedly sympathetic waiter standing in the middle of the restaurant conducting a mob farewell:

'*ADIÓS, DON JUAN . . . ADIÓS! . . . VIVA EL AMOR! . . . VIVA CASANOVA-A-A-A!*'

* * * * * * * * * *

'You know, Ellie, maybe you were right about the big Spanish families after all,' I said at length on the drive home. 'That waiter has got me thinking about all that many-sons-to-look-after-you-when-you're-old stuff. Hmm – he's definitely got something there. I like it.'

Ellie maintained a fittingly pregnant silence.

'Yeah – why not stick with the When in Rome formula?' I persisted. 'Why not? I've already had all the ammunition – the snails, the *all-i-oli* and the *Hierbas*, so why waste it?' I gave her a lascivious leer. 'What say we hit the sack early tonight, and I'll have a nightcap of that wormy orange juice you squeezed this morning – just in case?'

Ellie uttered a derisive snort, but as bedtime duly came round, she began to yield inevitably to nature's primal temptation – spellbound, perhaps, by the enchanting smack of garlic, anise and orange on my breath.

'OK, Don Juan,' she moaned teasingly, 'let's see what Mallorcan snail power is all about.'

'I knew you'd come out of your shell eventually, *querida*,' I whispered hoarsely, moving closer as sensuously as was possible over the lumpy mattress. *'Bésame mucho.'*

Yes, the stuff was working! I was even whispering sweet nothings in Spanish . . . and so what if it was only an old song title? I was really doing the authentic Latin lover business, and I wasn't even trying yet.

But oh, cruellest of fates . . . unknown to me, my mystical increase in libido had coincided with an even greater upsurge in the natural appetite of the immediate woodworm population; and just as Ellie and I were about to conclude our research into the local folk beliefs, the old bed gave up the unequal struggle and collapsed in a creaking, twanging paroxysm, leaving us flailing helplessly in a cloud of sawdust and feathers amid a heap of crumbled wood on the bedroom floor.

'Wow!' Ellie gasped through the dust. 'That was really something!'

I nodded in breathless agreement while a legion of micro passion-killers marched off victoriously to their next edible theatre of war.

'But I'm warning you,' Ellie spluttered, 'if I ever catch you eating snails again, I'm leaving home!'

– FOUR –

ORANGE GROVE ANGST

There were four little fields on the farm, running in a gentle, almost imperceptible slope from the high old wall bordering the lane, *el camino*, down to the *torrente*, the 'stream' which marked our western boundary. Judging by its deep, walled sides, the *torrente* must have run full with flood water from the surrounding mountains in bygone times, but not even the oldest of our *vecinos* in the valley could remember it other than it was on that balmy winter's morning – a sleepy little canyon with perhaps a trickle of water running unseen under a thick mat of brambles intertwined with grapevines which had tumbled down from their rustic trellises along the edges of the adjoining fields.

An ancient well, the source of all the water on which our fruit trees depended, stood in a corner of our *torrente* field, almost hidden behind a particularly lush mandarin orange tree whose branches were laden with ripe, juicy fruit. I didn't know why, but it was certainly the healthiest-looking tree on the whole farm.

I leaned on the well's parapet, its stone worn smooth by the ropes and buckets of countless generations of water carriers, and looked down to the glistening surface of the water far below. The well shaft was about four feet wide, and to the eternal credit of the long-forgotten masons who had built it many centuries ago, the meticulously-crafted stonework was still as sound as the day they had completed it. But not so the superstructure. All that was left of the original water-raising machinery was a rusty iron bar which straddled the well on elevated oak bearings, the rotted remnants of two heavy wooden cogwheels, and a broken timber beam.

Although I would not have to start irrigating the trees until the onset of the dry season in three or four months time, Tomàs Ferrer had already explained that all that would have to be done to extract the water from the well would be to switch on the electric pump which was installed in a little stone shelter nearby. The watering could then be done by hose from take-off taps on the galvanised pipe which had been laid along one entire side of the farm from the well to the lane. I looked carefully at the broken-down remains above the wellshaft, and while it was impossible for me immediately to fathom how this old 'machine' had worked, it was obvious that the principle was a lot more arduous than the modern convenience technique of simply throwing an electric switch.

I was startled out of my ponderings by the sound of twigs cracking behind me, and I turned to see the frail, stooped figure of old Maria Bauzá hobbling towards me and swiping away furiously with a small mattock hoe at the weeds which she noticed growing under her lemon trees.

'*Buenos días*, Señora Bauzá,' I called.

'*Buenos,*' she replied, a trifle breathless. 'Oh, *madre mía*, I must rest for a moment.' She lowered herself stiffly into a sitting position on the drystone wall that separated our two farms. 'That is the trouble with tractors and all those modern contraptions – can't clear out the weeds right up to the base of a tree like we could with a donkey or a mule. Noisy, smelly things, tractors. The old days were better.'

'Yes, I'm sure you're right . . . I suppose. And as a matter of fact, I was just wondering how they got water out of this well in the old days – you know, before electricity and pumps and things.' I slapped my hand against one of the old broken cogwheels. 'I mean, how did this all work?'

'*Sí, ah sí.*' She was already breathing more easily and had clasped both hands over the end of her hoe handle to rest her weight on the implement in the way that comes naturally to the old fieldworker. '*Sí, señor*, I must tell Jaume, that son-in-law of mine, to take more time with his tractor when he is weeding round the trees. Every weed you leave in the ground takes food and water away from your trees and *plantas cultivadas* – your tomatoes, peppers, beans . . . everything.' She looked over at the mess of wild greenery that was still covering every square inch of the land on our farm, and shook her head dolefully.

I couldn't help feeling ashamed, although I genuinely hadn't had time to buy the necessary equipment to allow me to get on with the job. I still suspected that Tomàs Ferrer should really have had the work done prior to our taking over the farm, but that didn't matter now and it didn't stop me from feeling conscience-stricken in front of our fastidious old neighbour.

'Perhaps, Señora Bauzá, you or your son-in-law could advise me which type of tractor would be best, and where I might buy one,' I pleaded.

'The water was drawn up from that well by donkey power, of course – or by a mule, if you had one. That is how it was all done,' said old Maria, slipping effortlessly into her quaint little habit of ignoring your present question and answering your previous one. 'I can tell by the puzzled look on your face that you do not understand, *señor*,' she added, slightly impatiently.

'No . . . well, yes . . . but I was just asking about the type of tractor –'

'Look,' she interrupted, pointing her hoe towards the well, 'if you listen, I will explain everything. Do you not have wells in Scotland?' She didn't bother to wait for a reply. 'You see that big cogwheel – the one lying flat, *horizontal?*' Again there was no pause for my response. 'Well, that wooden beam coming out from its hub was originally long enough to reach out beyond the well, and your donkey or mule was yoked to it there. *Me entiende?*'

I nodded to indicate that I understood, but old Maria was paying no attention anyway.

'So your donkey walked round and round the well,' she continued, her face creasing into one of her fabulous five-tooth smiles, 'and sometimes you had to tie an apple on a stick in front of it to make it walk . . . if it was a very stupid donkey.' She dissolved into quiet cascades of wheezy laughter. 'Ah *sí*, those were the days.'

I thought I would grab the opportunity offered by this whimsical interlude to press for information on the subject of tractor buying. 'So is there a place in Andratx where I can buy a –'

'*Momentito!*' The old woman waved her finger at me. 'I have not finished. If you want to learn about the old days, you have to listen carefully and do not interrupt.' She made herself more comfortable on her stony perch.

'*Lo siento,*' I apologised. This was clearly going to be a lengthy lecture, but I was already learning one thing – it was no good being impatient on such occasions. Taking half a day to have a chat was the Mallorcan way, so there was no point in trying to fight it. Just flow with the tide, be *tranquilo*; that was the only way, and I was trying hard to conform.

Maria was resting easily on her hoe again and looked all set to continue her history lesson. 'So your donkey walked round and round the well, as I said.' She took a deep breath. 'So the beam turned the horizontal cogwheel which turned the perpendicular cogwheel on the end of the iron shaft that goes over the top of the well. The shaft revolved and there was a big wheel in the middle of it with jars tied one after the other to a long loop of rope that went all the way down to the bottom of the well. *Comprende?*'

I nodded and waited silently as Maria took a little hankie from her sleeve, blew her nose delicately and then proceeded to examine the contents meticulously and approvingly before stuffing the hankie back up her sleeve.

'I used to blow my nose with my thumb when I was young – before I was taught how to act like a lady,' she admitted. 'But I have always thought that it is cleaner to use your thumb. Better on the ground than up your sleeve, no?' The five-tooth grin appeared again, and it was obvious that old Maria was delighting in this opportunity to address a captive audience on her favourite subject of 'the old days'. A long silence ensued as the old woman savoured her private memories.

This appeared to be an opportune moment to try again to change the subject to tractors and where to buy them. 'Eh, excuse me, Señora Bauzá, but –'

'No, no, no!' She shook her finger at me again. 'Always in such a hurry, you young people. I will explain everything, if you have patience.'

Maria pulled herself to her feet and shuffled carefully through a gap in the wall, vigorously resisting my courteous offers of assistance as she picked her way over the fallen stones. Tapping her hoe handle against a length of old metal guttering which ran from above the well to a large stone storage tank by the wall, she continued, 'So, as the jars went over the top of the wheel, the water which had been lifted up in them fell into this channel – it used to be made of wood and tiles, I seem to recall – and it poured into that *cisterna* over there where it was stored until it was needed. You can only take so much water out of a well at one time, you understand; then you must leave it to rest until the level rises up again.'

'And what about –'

'Then, when it was time to water the trees, you took your donkey and your plough and cut furrows along every row of trees in every field – all the furrows connecting up so that, when you pulled the bung out of the bottom of the *cisterna*, you could direct the water to all the trees on the farm in turn.'

'And all that by donkey power,' I remarked when I was sure she was finished.

'*Sí*, and these old, donkey-powered water wheels were called *norias* or *sínias* – brought to us in ancient times by the Arabs, they say. *Sí*, a useful little animal, a donkey. Much better for a farm than a tractor, you know.' Señora Bauzá sidled over to me and nudged my arm. 'Did you ever hear of diesel oil helping a tree to grow? No, of course not. *Bueno* – just take a look at that *mandarina* tree in front of the well. Is that not the healthiest tree on your whole farm? Of course it is!' She guided me by the elbow over to the tree. 'And do you know why this is the healthiest tree on the whole farm? No, of course not. Well, I can tell you, *señor*, that it is because they tied the donkey to this tree for a rest and a feed when it was working at the well. *Comprende?*'

'No, I . . . I'm afraid you've got me there . . .'

Old Maria looked up at me with a rueful shake of her head. '*Hombre*, the donkey stood under the tree and ate its food and drank its water. *Comprende?* So what would happen next? *Madre de Dios!* What goes in one end must come out of the other, no?'

'Oh, right! I see what you mean now,' I laughed stupidly. 'The donkey . . . er . . . it . . .'

'*Sí, claro!* Right here, under this tree – *aquí mismo!*' She thumped her hoe on the ground as if to ram the message home to my dense *extranjero* brain. Then, cupping one of the plump little oranges in her hand and examining it carefully, she added, 'Well, old Paco always claimed that was why this tree was so good, but now I am not so sure.'

'No?'

She raised her round shoulders into an exaggerated shrug and flashed her five teeth in a roguish little smile. 'Not unless it was a magic donkey, *señor*, because this tree is just as healthy as ever today, and it has been nearly twenty years since a donkey had a shit under it.'

The orchard positively rang with the tinkling cackle of her laughter.

'*Hola!* I hope Mama is not being a nuisance, *señor*.'

I looked in the direction of the voice and saw the rotund frame of a tall, elderly man standing on the other side of the wall in Señora Bauzá's lemon grove. A mop of white curly hair crowned his happy, avuncular features, and a pair of splendid horn-rimmed spectacles sat precariously on the end of his nose.

'Ah, Jaume, there you are,' piped Maria. 'Do you know that you left some weeds in that field again? You should sell that tractor of yours and get a nice *burro*.'

'*Sí, sí*, Mama – whatever you say,' he grinned, turning towards me. '*Qué tal, señor?* I am Jaume. And please do not believe everything my mother-in-law tells you. An old rascal, her.'

Maria muttered something in *mallorquín* that I was glad I couldn't understand.

'*Buenos días*, Jaume. My name is Peter. I come from Scotland,' I said as the big fellow shook my hand heartily.

'Oho! Peter is Pedro, no? So you are Don Pedro – Don Pedro de *Escocia, sí?*' Jaume could see that I liked the sound of that title, and his shoulders shook in silent laughter. He gave me a friendly but slightly too robust slap on the back, causing me to go into a little spasm of coughing and choking. Maria tittered delightedly. Jaume gave me another slap on the back.

'Well . . . I . . . I'm very pleased to meet you at last, Jaume,' I spluttered, flicking a tear from the corner of my eye as nonchalantly as I could. 'As you can see, our oranges are ready for picking, and Señora Bauzá mentioned before that you might be able to advise me how best to sell them.'

Old Maria immediately pushed the side of her hoe handle firmly against his chest, as if to halt any words before they could reach his mouth. 'He needs a tractor,' she announced decisively, reverting without warning to my earlier unanswered query. 'Well, *he* thinks he needs a tractor. *I* know he needs a *burro*, but if he wants one of those noisy, smelly contraptions like the rest of you modern *locos*, I suppose you had better tell him where to get one.' She tottered off towards the hole in the wall, mumbling in *mallorquín*.

I guessed that she was more than a little peeved that her exclusive dissertation on the old days had been interrupted by Jaume's arrival.

Her son-in-law chuckled good-naturedly and watched the old woman until she was safely back inside her own field. 'Do not worry, *señor*. I will be happy to try to answer *both* of your questions,' he said reassuringly. 'I hope Mama does not confuse you too much. She thinks in leapfrogs these days . . . a little difficult to follow until you are used to it.'

'Yes, but she's a real character, and I think her little quirks are quite charming. In fact, I'm becoming very fond of your mother-in-law. She's a wonderful old lady.'

Jaume smiled and added softly, 'She is also tough as an old terrier. Still comes up here from the village with me every morning and does her little bits of work on the *finca*. And do you know how old she is?'

'Yes, she told me herself the first time we met. Eighty-two, I think she said.'

Jaume laughed out loud. 'Ninety-two, more like. But she is still crafty enough to lie about her age. Typical woman, no?' He then cleared his throat and adopted a business-like demeanour. '*Bueno, señor* – now I will attempt to answer your questions. But I must warn you – I am no expert on farming.' He looked over the wall to see if old Maria was out of earshot and confided in a low voice, 'To be honest, I do not even like farming. Too much like hard work.'

We ambled off between the lines of trees and Jaume began to whistle quietly to himself. He didn't seem in any hurry to start giving me the advice I so badly needed. Indeed, he wasn't even bothering to look at the condition of the oranges or the weedy state of the land – or so I thought. After a couple of minutes, he wheeled round and we wandered back towards the well.

'No, I do not really like farming, *señor*,' he repeated at last. 'I only work the farm to keep my wife's mother happy. The farm is her life. She has been a widow for many years, you see, and the farm is where her memories are.' Jaume brushed a wisp of straw off his old woollen cardigan and tugged at its sides until he was just able to fasten the distressed garment over his corpulent midriff by the one remaining button. He looked at me solemnly over the top of his glasses. 'I am a waiter by profession, you know. Never a farmer . . . a waiter, me.' He adjusted his cardigan again, and his hand rose automatically to where his bow tie would once have been.

'Oh, a waiter, really? In a restaurant near here?' I asked innocently.

Jaume looked at me aghast. *'Restaurante, señor? Restaurante?'* He hitched up his baggy working trousers indignantly and jerked down the bottom of his over-stretched cardigan in a vain effort at making the two items of clothing meet in the middle. 'I was a waiter at the Hotel Son Vida – Mallorca's best, five-star grand-luxe! *Sí, sí, sí.'* He stuck out his chest proudly, and his trousers slipped back down below his paunch again. 'Oh, *sí, sí, sí, señor.* Hotel Son Vida for twenty-five years, me.'

'Well, that must have been a splendid career . . . to follow your profession in such a fine hotel. I can understand that you must feel very proud,' I said, without wishing to sound in any way patronising, but immediately fearing that I probably had. Jaume appeared to accept my compliment unreservedly, however.

'So you can appreciate, *señor*, that farming does not come naturally to a man like me.' He kicked at a weed. 'I am accustomed to the finer things in life.'

'I see. So you're a city man yourself? You're a native of Palma?'

'No . . . no,' Jaume replied, and I detected a slight reluctance in his tone. 'No, *señor* . . . I was actually born here in this little valley – in that old house with the sundial on the wall, near the top of the lane.'

'Oh, so your roots are really in the country then?'

'My roots may well be in the country, *señor*, but my heart is still in the Hotel Son Vida. They only retired me off early because my legs gave out. It happens in my *profesión*, you know. Ah *sí* – the legs and feet go.'

'But I suppose there are compensations, Jaume. It can't be too bad to retire to such a lovely place as this, particularly when the valley is really your home.'

'Perhaps,' he said wistfully, 'but I miss the busy atmosphere of the big hotel – the luxurious surroundings, the wealthy clients, the royalty. It is not easy to live the lonely life of a poor *paisano* after all that.' He gave a little chuckle. 'Why, the only clients I wait on now are Mama's hens and her pig . . . and they do not tip so generously as the guests at Son Vida.' Jaume clapped his hands to his ample belly and laughed jovially – like Father Christmas without the whiskers and red suit, I mused.

Despite his alleged pining for the glitzy hotel life, I guessed that old Jaume was about as contented as any man deserved to be, even if he didn't fully realise it himself. No doubt, it would be difficult living on a meagre income from the little farm after having been accustomed to the security of a regular wage, but at least he had had the family farm to return to in his old age – a lot more than most of his *profesión* could even wish for.

'Of course, I do get a small extra pension for my bad legs,' said Jaume, as if reading my thoughts, 'so I do not have to rely on making a living from the farm, *gracias a Dios*. No, I only work on the farm to keep Mama happy. Someday my wife and I will sell it, I suppose – after Mama has gone . . . if she ever goes.' He chuckled again. 'We prefer to live in our modern *apartamento* in the town, anyway. The old house along there – the one with the sundial where I was born – well, that was left to me when my father died, but we did not want to live in it . . . too old-fashioned. No, I like a modern *apartamento*, me . . . in Andratx where at least there is a little life.'

'But the old house where you were born . . . it's so beautiful. Ellie and I admire it every time we pass by. It's just so perfect . . . the little patio with the grapevines and all the geraniums, the well by the gate, and that old fig tree shading the doorway . . . and that, that amazing sundial on the wall. I mean, it's just like a picture postcard. It would be worth a fortune today. Don't you ever regret selling it?'

Jaume looked surprised. 'Regret selling it? Oh no, *señor*, I did not sell it. That house is for me to hand on to *my* son, you understand.' His face lit up. 'Yes, my son, José. He has two children of his own now – two little girls – *las preciosas*. José is in the army, you know. Doing very well, too. Got promotion this year and . . . and . . .' A look of sadness spread over Jaume's face and his voice dropped almost to a whisper. 'And his mother and I worry about them all the time. They sent him to the Peninsula when they promoted him – to the Basque Country . . . to Bilbao. It is very dangerous there for the army and their families. ETA, you know, with their bombs and their guns. We worry. We worry. And, of course . . . we miss them.' Jaume pushed his glasses up to the bridge of his nose nervously and he forced out a little cough. 'José is our only son, *señor* . . . our only son.'

The old man stood silently with his thoughts for a few moments, looking through glazed eyes over the trees to the far mountains. I couldn't help but think that, had this still been Señora Bauzá's 'good old days', Jaume and his son would have been working happily together on their little farm in this peaceful valley with no knowledge or desire for what Jaume now described as 'the finer things in life'.

'The price we pay for progress, Jaume,' I eventually said, thinking aloud.

'*Lo siento, señor*. I was miles away. I must apologise.' Jaume took a handkerchief from his pocket and gave his nose a good, loud blow. His glasses slipped down to the end of his nose again, restoring the jolly appearance to his face. 'No, I will not sell the old house, although I do get many offers. Ah *sí*, I get offers all the time from *extranjeros*, but it is better just to rent it out. It gives me a little extra on top of the pension, you know . . . and some day it will be José's.'

'Your son's a very lucky chap.'

'No, no. *Es normal*. The father passes on his house to his son. *Es absolutamente normal*.' Jaume placed a hand on my shoulder and ushered me off on another stroll through the orange grove. 'Of course, there are those two houses on Mama's farm – over there on top of that little hill by the Capdella road. Can you see them?' He pointed through the trees to two semi-ruined stone houses standing side by side looking down the valley towards the sea.

'Oh yes. We can see them from our upstairs windows. They're in a super spot. Tremendous views from there, I'm sure. Ellie and I have often said they would be fabulous houses if they were restored.'

'Do you think so, *señor*? Do you really think that is so?'

'I'm sure of it. They're just what all the foreigners are looking for – old stone houses to renovate – it's what they all dream about.'

'Mm-mm. So what do you think old houses like that would fetch? You see, Mama has given them to my wife and we have no use for them, as I have explained. But I did not think they would be worth much – all falling down and everything.'

'I really wouldn't know,' I said, shaking my head, 'but there are plenty of estate agents around who would value them for you.'

Jaume paused and looked me in the eye over the top of his glasses. 'But just as a matter of interest – if you were buying them, what would be a fair price . . . eh, in your opinion?'

'Well, that's a difficult one. I'd have to have a good look round them, but I suppose . . . at a very rough guess . . .' I named a figure that I thought, if anything, was an ambitious asking price.

'Each?' asked Jaume, wide-eyed and incredulous.

'Oh yes, each – naturally,' I assured him.

'Wrong, *señor*. *Incorrecto*. I have already been offered twice that amount – on several occasions – from Germans mainly. But no, I do not think I will sell them just now – not yet. *Un día*, perhaps. We shall see. *Vamos a ver*.'

I was dumbstruck. Here was this scruffily-dressed old fellow who looked as if he was down to his last few pesetas, while in reality he was a man of property, worth anything from half a million pounds upwards, *and* he was getting his retirement pension, plus his bad-legs allowance, plus the rent for the house with the sundial, plus the earnings from the farm. And he said he missed working as a waiter!

Yet Jaume was only one of countless elderly Mallorcans who chose to live the life of 'poor *paisanos*' whilst sitting on the potential wealth of inherited 'peasant' properties which had been worth almost nothing until the influx of *extranjeros* seeking peaceful retreats in rural Mallorca had pushed up values beyond their wildest dreams. And the prices of their long-deserted stone cottages just kept rising and rising. The old Mallorcans saw to that. They were canny country folk; they were small farmers, so they all knew how to milk. And why not? As one old *campesino* told me, cows with golden tits are about as easy to find as lumps of unicorn dung, so if such a cow should chance to wander into your byre, you must milk her dry.

There was no doubt that Jaume also subscribed to this simple pastoral philosophy, and his bucket was at the ready.

'But enough of my little problems. You want to know about selling your fruit,' he said, stopping beside a decidedly poorly-looking orange tree. 'Do you know much about oranges, *señor?*'

'I know absolutely nothing about oranges, or any other kind of fruit for that matter,' I confessed.

'Well, you were right when you said that your oranges are ready to pick. Most of them are, and the others will come on during the next few months – as the yellow ripens out of their skins, you know. But do not pick them just because they look ready. They will keep better on the trees until you know that you have a buyer waiting for them.'

'That's just the problem, Jaume. I don't have a buyer for them and I don't know where to start looking for one. All Tomàs Ferrer told me was that old Paco used to sell most of this farm's fruit from a stall at Andratx market on Wednesdays, but I won't be allowed a licence to do that until I get my final resident's permit from the police, and that'll take months.'

Jaume stuck out his bottom lip pensively. 'Hmmm. *Es cierto*. You do have a problem – *un problema grande*. I do not grow many oranges, me – mostly lemons. And I sell them direct to the little grocery stores, *las colmadas*, in Andratx and round about there. I could charge a bit more if I had a stall at the market too, but it is easier just to deliver the fruit to the shops, so that is what I have been doing for years.'

'Right, so maybe I could do that with the oranges, eh?' I asked hopefully.

'*Posiblemente, señor* – but I do not think so. All the orange-growers around here already have the best outlets tied up. You can ask at the *colmadas*, of course, but . . .' He shook his head gravely.

'Well, I can't just let them rot on the trees. I've got to sell them somehow. Surely there are wholesalers who will give me some sort of price for them. Anything.' I was beginning to feel helpless, alien, and right out of my depth. What the hell had possessed me to get into a business that I knew nothing about, *and* in a foreign country where I couldn't even speak the language adequately? After all, I certainly wouldn't have got involved in the fruit business back home. No way!

Jaume laughed benignly and gave me a couple of little pats on the arm. 'Do not look so worried, *señor*. *Usted tranquilo*. In Mallorca, everything is possible.'

It was all right for him, I thought to myself. He could afford to laugh his benign laughs. He was a native here and he had it made. The old bugger was loaded, but I was fifteen hundred miles from home, I had hitched my whole caboodle to this Mallorcan fruit cart, and already the wheels were falling off.

Jaume started to whistle again. He pulled an old envelope and a stubby pencil from his shirt pocket and scribbled something down. 'Here you are, *señor*. You take this. It is the name of a fruit merchant along the coast in Peguera. He runs a small business in a lock-up behind the Bar El Piano. If ever I have any surplus lemons, I take them to him. He is very fair, *muy razonable*. He will give you an acceptable price, I am sure . . . eh, until you find more profitable outlets of your own, of course.'

'*Señor Jeronimo – Frutas Frescas*' read Jaume's note. I felt as if I had just found a fiver in Sauchiehall Street on a Saturday night. I'd been saved by a miracle. I shook Jaume's hand and thanked him profusely. How could I ever repay him for giving me my first business break in Mallorca? He was a good neighbour and a real gentleman – *un . . . un caballero*.

But my elation was to be short-lived.

'Unfortunately, *señor*, you have diseases on some of your trees,' droned Jaume woefully, peering through his spectacles at the curled leaves and stunted fruit on the adjacent tree. 'Look . . . this black sticky stuff on the oranges.'

'Yes, I've noticed that. Not too good, eh?' I waited anxiously while Jaume whistled another measure as an accompaniment to his closer examination of the problem.

'A bug, *un parásito*, I think. Mm-mm, but I cannot say for sure. You need to ask an expert, *señor*.' Jaume pushed his glasses back up from the end of his nose. 'Pepe Suau – he is your man. Does all my trees for me, him – spraying, pruning, everything. A true *maestro de los árboles*. Mallorca's best.'

Now we were getting somewhere. I had finally found a real tree expert. 'And where can I find him, this Pep Su-whatshisname?' I asked urgently.

'Pepe Suau? Oh, you do not have to find him. He will be here in the spring as usual – over at my place *como siempre*.' Jaume chortled soothingly, like some massive clucking hen. '*Usted tranquilo, señor*. Do not get excited. This is Mallorca. *Todo va bien*. Everything works out well . . . in time. Just give it time.'

It seemed I still had a lot of relaxing to do before I got my approach to life and its problems right. I would have to work much harder at slowing down. '*Siempre Paciencia*', patience at all times, would have to be my motto if I was ever to integrate properly into the lifestyle and pace of rural Spain. I still had a long way to go to purge myself of all the absurdities of the Anglo-Saxon work ethic. *Hombre*, you only work to live, not the other way round – that was the local perspective, and who was I to argue?

Jaume had gone over to an extremely sorry-looking tree with no leaves and with ugly spills of resin oozing out of its bark. He gave its trunk a gentle push, and a large branch fell off.

'It would appear that this old plum tree is dead then, Jaume?' I said apprehensively, my northern tensions rising again.

'*Casi casi*. Almost, but maybe not completely. As you know, I am not an expert. But never fear – if anyone can save it, Pepe Suau will. Oh, and it is an apricot tree, *señor*. That is a plum tree there,' he smiled, pointing at an equally forlorn specimen. 'Your kakis look quite good, though.'

'My *kakis*?' I looked at him blankly.

'Persimmon, you call them in English, I think,' laughed Jaume.

I was none the wiser.

'But look over there. See how overgrown your fig trees are. And the quinces, the pomegranates – all of them. Mm-hmm, there will be much work here for Pepe Suau. *Mucho, mucho, mucho.*'

Jaume casually whistled his way over to a lemon tree, and my tensions were now escalating into panic – full-scale, out-of-his-depth, drowning man's panic. The Ferrers had sold us a real pig in a poke. Putting this dump right would cost a fortune. I'd make nothing out of it. I'd be flat broke long before we could ever see a return on our investment. All our money gone. I'd be a bankrupt in Spain – and a foreign one to boot. Christ – they'd throw me in jail! Ellie and the boys would be destitute!

'You see these long, green shoots growing up from the main branches?' said Jaume at the lemon tree. 'Well, these are suckers.'

'Don't mention the word "sucker" to me, please,' I said under my breath.

'They take all the good out of the tree,' he continued. 'All of your citrus trees are full of them. They will all have to be cut out. But never fear – Pepe will do all that for you in the spring. Mm-hmm. Ah *sí, amigo,* Pepe will have much work to do here, because he will also have to . . .' Somehow sensing that I had had enough bad news for one day, Jaume looked round at me and smiled understandingly. He put a fatherly arm round my shoulder and announced breezily, '*Nos vamos,* Don Pedro. The trees will wait till spring. We will go now and see my tractor. You want to know about Spanish tractors, no?'

'Yes, well . . . you can see the weeds . . . it's something else I have to do and . . . well, I don't even know where to . . .' I was descending into a pit of despair.

'*Tranquilo, tranquilo, siempre tranquilo,* Don Pedro.' Old Jaume was almost singing the words as he patted my shoulder soothingly and guided me towards the hole in the wall. 'As I used to say to little José when he was scared at night – things will be better in the morning. And, Pedro, things will be better for you in the spring, believe me.'

This kindly old man who had met me for the first time only half an hour earlier was already taking me under his wing – letting me know that he understood and cared. I began to realise just how fortunate we were to have been blessed with such good neighbours as Jaume and old Maria.

'Yoo-hoo-oo! Yoo-hoo-oo!' It was Ellie, basket and secateurs in hand, emerging from the green depths of an orange tree in the next field.

I beckoned her over. 'Ellie, this is Señora Bauzá's son-in-law, Jaume. He's taking me over to their *finca* to see his tractor.'

'I'm very pleased to meet you,' smiled Ellie, offering Jaume a sticky, orange-tainted hand.

'*Encantado,*' replied Jaume with a little bow. He then stood grinning shyly with one arm folded behind his back, almost as if he were waiting for Ellie to order a coffee.

'Would you mind if I tagged along?' she asked instead.

Jaume nodded his head politely and replied in perfect English, 'This way please, madame.'

'Hey, I didn't realise you could speak English,' I said, totally surprised.

Jaume led us off in the direction of his *finca*, laughing coyly. 'Oh, I speak French, German and even Italian also, but all very badly. I only know the things I needed to say as a waiter in the hotel.'

'Maybe so,' I replied, 'but it'll be a great help to me if we can sometimes talk together in English, because –'

'Oh no, no. That would not help you at all,' interrupted Jaume, stopping and turning round to look at me over the top of his glasses like some wise old owl. 'This is my country and you must speak to me in my language. You can practise your Spanish on me whenever you like. *Siempre español*, Pedro. *Siempre español*.'

Duly admonished, we followed our paternal neighbour in single file along a little track at the edge of one of his fields, passing by rows of lemon trees full of large, shiny fruit which made our inferior examples seem like a different species entirely.

'Not too bad, the lemons, no?' said Jaume, appearing almost to read my thoughts again. 'Yours will be like that also. Just leave everything to Pepe Suau. *No problemas*.'

We soon reached a clearing in the trees, and there, at the far side, was the nerve-centre of the Bauzá farm – a higgledy-piggledy shambles of little sheds, shacks and pens made from as diverse an assortment of materials as could be found in any builders' merchant's – or junk yard. At the centre, flanked by two bushy palm trees, stood what looked like the original *casita de aperos*, a little stone cabin for storing farm tools and implements. A ramshackle framework of rusty conduit pipes had been built along the front of the *casita* and was bound together by a wild weave of grapevines and bougainvillaea, covering the makeshift pergola and providing shade above the little table and two wooden benches that were standing outside the low front door.

The *casita de aperos* had been extended on one side with bits of old telegraph poles supporting concrete beams which in turn bore the weight of a low, undulating roof of miscellaneous timber joists and some corrugated asbestos sheets. Rocks and lumps of broken breezeblock had been laid randomly on the roof, presumably to discourage it from migrating southwards when the *Tramuntana* gales blew through. Jaume's old Citroën Dyane car sat snugly inside this open-fronted, free-form shed, surrounded by a glorious chaos of fruit crates, straw bales and bundles of cane.

At the other side of the *casita* were the quarters of the farm animals, *las dependencias de los animales*. A dilapidated lean-to hut of Mallorcan sandstone, with a roof of ancient terracotta tiles to match those on the *casita*, rested comfortably against the *finca*'s perimeter wall which ran along the back of the little farmstead. Adjoining this, a row of small corrals had been constructed within a low, uneven enclosure of hollow concrete blocks, staked here and there with crooked posts of dead almond branches to which lengths of chicken wire had been tied. The gates to each pen were old metal bed ends – one still sporting a solitary brass knob.

Jaume had hurried on ahead to announce our arrival to old Maria and his wife, so Ellie and I paused at the edge of the clearing to savour the sight of this curious fusion of traditional and contemporary architectural disorder. Somehow, it had a rustic charm all of its own – tumbledown and shabby, yet perfectly at one with the majestic Mallorcan scenery surrounding it. This tiny huddle of a farmyard, nestling in its secret sunny corner of a lemon grove beneath the spectacular backdrop of towering mountains, created a remarkable picture of unrefined, idyllic beauty.

No sooner had old Maria been made aware of our impending visit than she brushed Jaume and her daughter aside and shuffled over to greet us as fast as her stiff old pins would carry her. *'Ah, la señora de Escocia. Qué bellísima!'* she exclaimed, revealing her quintet of tombstone teeth in a huge smile of welcome as she grabbed Ellie by the back of the neck and pulled her head down to cheek-kissing level. *'Bien venido a mi finca.'* She took Ellie's arm and chattered away to her non-stop until we reached the *casita*, where she invited Ellie to sit with her at the little table. Meanwhile, I noted that I was being ignored completely by the old woman, who was presumably regarding me as persona non grata for daring to speak to Jaume before she had completed her lecture to me back at the well. She was now totally absorbed in Ellie and her tongue never lay still for a second.

Jaume waited patiently till Maria paused to draw breath, then he introduced us to his wife, Antonia – a neat little woman with a pretty face and a ready smile. She was the ideal foil for Jaume, and together they made a handsome and homely couple.

'Bueno, Pedro. If you are ready, we shall go to see the tractor,' suggested Jaume without further ado.

'Tractor? Bah! Noisy, smelly thing,' scoffed his mother-in-law. 'Come *señora* – let me show you my animals. I have some good hens just now. Did you like the eggs I gave you? Yes, of course you did. Well, I shall give you some more today. Did you know that Señora Ferrer strangled all the hens on your farm for her pot before you arrived? You must get some new hens, you know – and a pig. I will show you my pig.' She shoved Ellie gently in the direction of the first pen and took her arm again. 'Do you want to come over when we kill the pig, *señora*? Yes, of course you do. That will be a real *fiesta*, just like the old days, and I can show you how to make Mallorcan sausages . . . *camaiot, sobrasada, butifarra*. We use the pig's blood for the *butifarras*, you know.'

Jaume and his wife looked at each other, shrugged, then shook their heads resignedly.

'I have to apologise, *señor*,' said Antonia. 'I hope your wife does not mind being dragged round to see the animals. My mother does not mean to be selfish, but she is very proud of her *finca* and, well – she does not meet many new people, you understand.'

'Yes, I understand, and please don't worry. Ellie's delighted to be shown around. Like me, she has a lot to learn.'

'Ah, *gracias*, *señor*. You are very kind.' She smiled sweetly, then turned to Jaume. 'When you come back from the tractor, we must welcome our new neighbours properly. They must think us very rude.'

Jaume grinned knowingly and patted his stomach. 'I think my wife has a little surprise in store for you, Pedro. Mm-hmm.'

To get to the tractor, Jaume first had to take his car out of the shed, much to the annoyance of a pair of ducks who had been sunbathing in front of it, and to the shock of a hen which fluttered down in a cacophony of bad-tempered cackling from the car's canvas roof, where it had been enjoying a quiet nap.

'This type of tractor is best for here – best for this kind of soil – best for getting under the branches of the trees. It is the best, Pedro – the best,' pronounced Jaume as he removed the tarpaulin and shook off the hen droppings. He looked at his tractor adoringly. *'Buena máquina, no?'*

I was stuck for words. I knew that the tractors used on these little fruit farms were small in comparison with the huge brutes we were accustomed to back home, but this, this tractor of Jaume's . . . dammit, this wasn't a tractor. It was a motorised pram!

Jaume mistook my expression of surprised disbelief for one of open-mouthed admiration. 'Aha . . . aha. I knew you would be impressed, Pedro. It is a real beauty, no? A *Barbieri* . . . from *Italia!*' He beamed proudly and stood aside to give me an unobstructed view of his pride and joy.

The first thing that troubled me was that this 'tractor' only had two wheels. They were sturdy little wheels all right, with proper knobbly tractor tyres, but there were only two of them – one on either side of the little engine. Secondly, it had no seat, and thirdly, it had no steering wheel – only a pair of long handlebars with levers and things . . . almost like the controls of a vintage motor bike.

'Well, it really is a beauty, Jaume,' I lied diplomatically. 'It's . . . it's red and . . . and white . . . yes, red and white . . . nice colours.'

'No, no, no, no, no!' Jaume wagged his forefinger in my face. 'The colour does not matter. *No importa!* What matters is the power, and this tractor has the power – *doce caballos!*' He clenched his fist to emphasise the point.

'Doce caballos – twelve horse-power, eh? Well, well.' I didn't think it would be prudent or even relevant to tell him that I was more used to tractors with engines ten times that powerful, and with four wheels, and a seat, and an air-

conditioned cab. No, his was the right type of tractor for this type of farm, so I had better just accept the fact and listen to his advice. I needed it.

'The other thing, Pedro, is that this tractor has a diesel engine.' This time, he clenched both fists for maximum effect. 'Diesel . . . you must have a tractor with a diesel engine. Two-stroke petrol engines? No use. *Basura!*'

'Not enough power from the two-stroke petrol engines, correct?'

'*Correcto. Hombre*, the land here on the bottom of this valley is so heavy. It is muddy and sticky when it is wet, and it is like concrete when it is dry. You need the power of a diesel tractor to cultivate it.'

'OK, Jaume, I understand that. But tell me – er, ehm . . . four wheels. I've seen tractors here with four wheels, so I just wondered –'

'No, Pedro, no.' Jaume waved all five fingers in my face. 'Listen, if you have four wheels on a tractor, you also need to have a seat on the tractor, no?'

'Yes, that's right, and I just thought it would be better to –'

Jaume laid his hand heavily on my shoulder. 'Just think about it. Think about all those branches on your trees that almost touch the ground. The tips of some of those branches are a good two metres out from the tree trunks, no?'

'*Sí.*'

'*Bueno*. If you are sitting on a tractor with a seat, you cannot duck under those low branches, so you either drive on and break the branches – and maybe your face too – or you steer round and leave the weeds under the low branches, then go back and take the weeds out by hand later. But that is too much work, Pedro. *Demasiado trabajo, no?*'

'*Sí.*'

'*Bueno*. So you walk behind a two-wheel tractor like this one and you can dodge anything. Even I can do it with my bad legs. *Hombre*, a two-wheel tractor is the next best thing to a donkey. *Incorrecto* – it is better, because it has more power than a donkey *and* it will not shit on your boots.'

Jaume had convinced me. I needed a two-wheel tractor, *doce caballos – mínimo*.

He draped the tarpaulin lovingly over the tiny machine. '*Ahora*, Pedro, we must rescue your *esposa* from Mama. *Nos vamos.*'

We could hear the old woman's voice coming from the direction of the little lean-to shack on the other side of the *casita*. 'So you see, *señora*, do not believe anyone who tells you that ruffled feathers round the hen's vent are a sure sign that the hen pays frequent visits to the laying box. No, no, no. Look – come into the sunlight and I will show you.'

Ellie appeared from the shack, stooping carefully under the low lintel. Old Maria followed closely behind, holding a protesting hen upside-down by the legs. She poked a finger into its nether parts and the hen squawked a futile objection, cocking one eye sideways at its captor and blinking indignantly.

'There,' said the old woman triumphantly, holding the flapping bird up to Ellie's face. 'That is what I call the best sign of a good-laying hen – a nice clean arsehole. *Bellísima!*'

'Mama!' called Jaume's wife from the doorway of the *casita*. 'That is enough now. I am sure the *señora* knows all about hens without getting lessons from you. *Ven aquí*. Come and sit down. I must welcome our new neighbours properly.'

Maria huffed a comment in *mallorquín* and unceremoniously chucked the hen back inside the shack.

Once Jaume had seen to it that we were seated comfortably at the little table, Antonia went inside the *casita* and returned with a large loaf of homemade bread, a dish of bright red tomatoes and an *aceitera*, a small tin container for olive oil, resembling a miniature watering can with a long, thin spout.

'If I had known that *los señores* were coming to visit, I would have brought something more *especial* from my kitchen at the *apartamento* in Andratx, but I think – I hope – you will enjoy this traditional Mallorcan country snack,' said Antonia, cutting the fresh, crusty bread into thick slices.

I noticed that Jaume had meanwhile slipped off quietly in the direction of the tractor shed.

'I picked these this very morning,' chipped in Maria as her daughter cut some of the juicy tomatoes in half. 'Look, you can see them growing over there between those almond trees. See the cane stakes? If the weather stays fine, we will eat them fresh from the field until Christmas, then I will gather what is left, tie them to strings by their stalks, and hang them up indoors. Those *tomates de ramillete* will keep us going until next year's crop is ready . . . just like the old days.'

'Of course, nowadays we bottle some of the surplus tomatoes also,' added Antonia as she rubbed the cut face of the tomatoes into each slice of bread until the surface glowed pink.

Old Maria tapped the back of Ellie's hand to attract her attention. 'These days, we bottle some of the tomatoes too, you know.'

'Really? Well that's a good idea,' said Ellie, trying to sound surprised.

'Oh, *sí*, *sí*. *Una idea maravillosa*,' replied the old woman, winking an eye. 'I like to keep up with the times.' She tapped Ellie's hand again. 'You must grow your own tomatoes next year, *señora*. You will not need many – only a row or two of them in a nice sheltered spot. But be sure to plant them in the waxing moon.'

'Uh-huh, *sí*,' said Ellie, totally baffled.

Antonia was now sprinkling the tomato-smeared bread with drops of purest olive oil from her *aceitera*. She then cut the remaining tomatoes into slices with which she neatly covered each slab of prepared bread. A fleeting shake of her olive wood saltcellar, and the treat was complete. '*Vale, señores*. Here we have the *Pa-amb-oli*. I hope you enjoy it.'

We were invited to help ourselves from the large white plate which she placed in the centre of the table.

'Ahem, Antonia,' said old Maria, taking on an air of authority, 'you should have explained that this is not just ordinary *Pa-amb-oli*. This is the special *Pa-amb-oli i tomàtiga* that you have prepared for our guests.' Turning to Ellie, she went on: 'That is because she has added the tomatoes, you understand. And, you know, I picked the tomatoes this very morning from –'

'*Sí, sí*, Mama – we know all about your tomatoes,' sighed her daughter, spooning some dark green, fleshy twigs from a pickle jar. 'Better to explain about these titbits, no?'

'Ah, the *fonoi marí* – a real Mallorcan delicacy. It is the sea fennel which grows on our rocky coasts, you know. We gather it up, steep it in brine, then pickle it in vinegar. *Madre mía*, no *Pa-amb-oli* is complete without some *fonoi marí* to go with it. And it costs nothing. Here – try some, *señora*. And, ehm, if you like it, I can show you where to find it.'

Ellie nibbled tentatively at the end of a sprig, then popped the whole piece in her mouth. 'Mm – delicious. Just like a little, knobbly gherkin, but tastier. Mmm!'

'*Bueno, no?*' enquired Maria, flashing her five teeth in an expectant grin.

'*Bueno, sí!*' Ellie replied fluently. 'Very *bueno!*'

'Ah, in the old days, we could eat like this all the time, *señora* – like queens. The *Pa-amb-oli* and the *fonoi marí* – and it cost nothing. We had everything – oil from the olives, *fonoi* from the rocks, tomatoes from the field, and all free. We only needed to bake the bread. In the old days, all the houses had a big oven in the stone wall, and when it was still hot from baking the bread, we used to roast a little pig – on fiestas and special occasions. *Mierda!* Talking of pigs . . .'

Jaume had reappeared on the scene, grinning broadly and carefully bearing a dusty glass flagon in both hands. 'The last *garrafa* of my own red wine of the year 1982,' he announced grandly, pulling the cork with a professional 'pop!' and pouring the ruby liquid with studied reverence into a selection of jamjars which his wife had fetched from the *casita*. 'Oho, the 1982 – what a vintage,' he drooled, sniffing a lungful of the bouquet from his 'glass'. He closed his eyes in ecstasy, coughed involuntarily, then stood smartly to attention to propose a toast: 'To our new neighbours! May your life in our valley be a happy one. *Molts d'anys!*'

Antonia and old Maria raised their jamjars and chanted in unison, '*A nuestros vecinos nuevos. Molts d'anys!*'

Jaume gulped down a manly draught of his treasured '82, his wife sipped the tiniest sip of it, while old Maria dumped hers back on the table and pushed it aside contemptuously, muttering in *mallorquín* about the 'piss of a donkey' and 'having no desire to swap her trusty hoe for a white stick just yet'.

Undaunted, I took a generous slug of Jaume's homemade *vino*, sloshed it around my mouth connoisseur-style, then swallowed the lot wino-style.

Jaume noted my reactions carefully over his horn-rims.

'Teeth still intact?' I asked myself, counting them with my tongue – just to make certain. Whew! I had tasted some so-called full-bodied wines in my time, but this stuff of Jaume's was the Incredible Hulk of viniculture.

'I can see that you, too, are an *aficionado* of fine wine, Pedro,' he said, managing to read something complimentary in my stunned expression. 'And you need say nothing. I know this '82 of mine is . . . well, there is no word to describe it adequately.'

'Frightening' might do for starters, I reasoned silently. This certainly wasn't boys' wine.

'And now you must sample my wife's *Pa-amb-oli*. I think you will find that it complements the wine beautifully,' advised Jaume, slipping effortlessly into waiter-spiel.

I had often heard people wax lyrical about this basic Mallorcan dish, but its name – meaning literally 'bread with oil' – had never evoked an appetising image for me. I hadn't realised until then what I had been missing. The first bite of this deceptively simple open sandwich seduced my taste buds: soft, sharp, creamy, sweet, crisp, crunchy, rich, bitter, fresh . . . fabulous. And, of course, Jaume was right. The *Pa-amb-oli* and his staunch *vino tinto* went together perfectly. Well, I thought so anyway, although I noticed that Ellie and Antonia had discreetly poured the contents of their glasses into those of their more discerning husbands.

The warm midday sun shone down through the gnarled filigree of vines above us, covering the table-top in a slowly shifting lacework of light. A little dandy of a bantam cockerel that had been strutting arrogantly around the yard since the arrival of the *Pa-amb-oli* suddenly flapped onto the table, let rip with a rather puny, broken-voiced 'Cock-a-doodle-doo', and tip-toed boldly towards the big white communal plate.

'Aha, General Franco,' smiled Antonia, sticking out her hand gently to bar the bird's path towards the food. 'In this *finca*, he thinks he is the boss, *el jefe*. He makes a lot of noise and likes to show off, but he is really a little gentleman.'

'Little gentleman, nothing,' scoffed old Maria, swiping the cockerel off the table with a single backwards flip of her hand. 'The last time he tried that trick, he took a beakful of bread, shat on the plate, then jumped down and mounted a nice little hen right there in front of me. Some gentleman.'

'Mama, *por favor*,' pleaded Antonia, visibly embarrassed.

'No, no, no,' continued her mother irately. 'General Franco or no General Franco, he will finish up with his neck stretched if I get a hold of him. The next time he lands on this table will be inside a pot.'

Jaume sat contentedly sipping his wine and munching his *Pa-amb-oli*, a peaceful smile on his face, and his stomach protruding defiantly between his trousers and his cardigan.

'Well, Jaume,' I said, already feeling the mellow warmth of his awe-inspiring 1982 *tinto* spreading through my veins, 'I haven't had any experience of your five-star, grand-luxe Hotel Son Vida, but if the setting, food and wine could match this, it would get a top rating in my book.'

'Oho, Pedro,' grinned Jaume modestly, 'you are too kind, too kind.' He poured another jamjarful of wine for me – though not quite so delicately this time.

I could hear that Ellie and Antonia were getting along famously, and the fact that they had no common language didn't seem to present any real handicap. Even the complication of old Maria's time-delayed contributions failed to slow the flow of conversation.

'About tractors, Pedro,' said Jaume, half turning his back on the chattering ladies. He tapped the side of his nose with his forefinger and winked confidentially. 'I think I know where you could buy one just like mine – a *Barbieri* – for a very good price.'

'*Sí?*'

'*Sí*. You know the workshop of Juan Juan, *el carpintero*, in Andratx?'

'No.'

'*Bueno*. It is easy to find – on a corner of the street between the market place and *el cine Argentina*. You will see all the broken window shutters, *las persianas*, lying outside his door waiting to be repaired.'

'*Sí?*'

'*Sí*. And I have heard that he bought a new *Barbieri* recently which he does not think is right for his *finca* up there in the mountains.'

'No?'

'No. Too steep up there for a two-wheeler. Now he wants a tractor with four wheels and a seat.' Jaume chortled jovially. 'But I think it is really the seat which is the attraction.'

'Well, well, that's good to know, Jaume. I needed to find a carpenter to mend some shutters anyway, so I'll go and see Señor Juan Juan and maybe kill two birds with one stone.'

'*Cómo? Por qué* you want to kill birds at the workshop of *el carpintero? No es posible.*' Jaume was clearly confused.

'It's all right, Jaume. *No importa* – just an expression in *Ingles.*' I decided that it would be easier to change the subject than to attempt an explanation. 'Ahem, I see you have some sheep in one of your pens over there, Jaume. Are sheep useful animals on a fruit farm?'

Jaume raised his shoulders. 'Useful?' His shoulders were raised a bit higher. 'Not very useful, no. I just keep a few sheep because they are more interesting than lemons, that is why.' He relaxed his shoulders and took another sip of wine. 'And it means that we usually have a lamb for the table, *naturalmente.*' The last of his *Pa-amb-oli* disappeared into his mouth, and he licked the olive oil and tomato juice slowly and deliberately off each finger in turn with a loud smack of his lips. '*Fabuloso!*'

'Yes, the *Pa-amb-oli* certainly was fabulous, Jaume. I'd been told about it many times, but I never believed it could be just so delicious.'

'Well, like so many good things in life – just a question of improving what nature has provided, Pedro.' He looked at me over his spectacles and tapped the side of his nose again. 'The trick is knowing *how* to improve what nature has provided, no?' The final dregs of his great vintage were unceremoniously shaken from the flagon into our glasses.

'Take this wine, for example – this *vino clásico*, Pedro. It is all the work of nature, of course, with only the expertise, the skill, the . . . the . . . the magic provided by me.'

'*Sí*, Jaume, *sí* . . . *sí*,' I replied falteringly, the noonday intake of half a litre of his formidable 1982 having the predictable effect on my thought and speech functions.

'Ah *sí*, Pedro . . . the gift of wine . . . the gift of wine.' Jaume took another slurp. 'I plant the vine, the vine draws the moisture from the soil and uses it to swell the grape. And then the grape . . . eh, what was I talking about? . . . *Sí*, then the grape is ripened by *el sol*. *Sí*, *el bon sol*. Mm-hmm, I pick the grapes, squeeze the juice from them, and *vale!* . . . the gift, the miraculous gift of wine.' He held his jamjar skywards in open admiration of what little remained of his most successful joint venture yet with Mother Nature.

He was shaken out of his reverie by a well-aimed crust of bread which hit him squarely on the side of the head, knocking his spectacles askew and doing nothing to enhance the image of master oenologist which he had been gamely trying to portray. The missile had been hurled from the other end of the table by old Maria, the accuracy of her shot perhaps suggesting a direct Bauzá lineage back to the legendary Balearic Slingers of Roman times.

'Of course sheep are useful animals on a fruit farm,' she shouted at her dishevelled son-in-law. 'Why do you not tell *el señor* how they eat the weeds off the land before you plough? *Madre mía*, he could do with some sheep over on his *finca* right now.' Maria then rejoined the on-going conversation of the women, albeit a topic behind.

'Ah *sí* . . . *correcto* . . . *absolutamente correcto*, Pedro,' flustered Jaume, fumbling with his specs. 'I was going to tell you about that, but . . . *el vino* . . . *el vino*. I, ehm . . . had to explain about *el vino* first. But Mama is correct, of course.'

I closed one eye and concentrated hard.

Jaume made a nervous adjustment to his imaginary bow tie. 'You know old Pep, your neighbour on the other side of the lane?'

'Well, I've seen him around, but I haven't met him yet.'

'Anyway, Pep has many sheep. A very big flock. The biggest flock of sheep in the valley. I think he has, oh . . . *posiblemente* twenty-five ewes – maybe twenty-six – and there are always some lambs, *naturalmente*. Well, old Pep is always looking for fresh grazing for his sheep. He grazes them on *fincas* all round the valley, you know.'

'Oh yes. I've often heard the sheep bells. Are those from his sheep?'

'*Sí, absolutamente*. He knows where every sheep is, even in the dark, when it is wearing a bell. Everyone is happy to have their weeds grazed by Pep's sheep before they plough. Nobody has so many sheep as Pep any more. In fact, on most *fincas*, there are no sheep at all.'

'So Pep would bring his sheep to eat our weeds?'

'*Sí*, if you ask him. He only has to drive them over the *camino* from his *finca* to yours. Nothing could be easier.' Jaume's expression turned serious. 'But, Pedro – I must tell you this. They will not be allowed to graze the field by the *torrente*, the field with the well.'

'No? Why not? It has the most weeds of all – covered in all that wild clover stuff.'

Jaume shook his head and frowned. 'No, no, Pedro – that is not clover. It is a weed called *Maria*. Some call it *vinagrella*, because it tastes of vinegar. It may look a bit like clover, but when it flowers, its flowers are yellow, so it is not clover. If the sheep eat too much of it, well . . .' He looked sideways at the ladies and confided quietly, 'It . . . it does not agree with the sheep's digestion.'

'That's good advice, Jaume. *Muchas gracias*. I'll remember about the *Maria* weed.'

'*No problemas*, Pedro. Old Pep will keep you right anyway. *Sí, sí, sí.*' Jaume leaned back and yawned luxuriantly.

Meanwhile, the ladies had left the table and were standing in the sunshine, huddled together in a tripartite ceremony of farewell hugs and kisses.

I stood up and turned to offer Jaume my thanks for his hospitality and advice, but he was already asleep – head bowed, chin on chest, spectacles teetering on the end of his nose, and his cardigan draped freely on either side of his bulging tummy. The button had finally conceded the inevitable and was lying in the upturned palm of his hand.

I thanked and said goodbye instead to his *señoras*, then Ellie and I set off homewards along the dusty dirt path beneath the lemon trees – Ellie with a basketful of fresh eggs, and I with a headful of little men with pneumatic drills.

'Ellie,' I said, 'I know it's going to take more time yet and . . . and . . . it'll be painful at times, but I think I'm really learning how to be *tranquilo* . . . at last.'

'Yes, yes, dear. And the next lesson in "*tranquilo*ness" is to come home and pop off to your bed for your first genuine *siesta*. I think old Jaume has taught you why those are so essential here too.'

'Dead right, Ellie. And I've learned a lot of other important things from Jaume today too . . . tractors, fruit merchants,

Pepe Su . . . Su . . . and sheep and everything. I'm a lot more *tranquilo* now all right.'

'Yes, yes, dear. And nothing to do with the private pre-lunch tasting of his vino collapso, I'm sure.'

There was no answer to that.

'And do not forget, *señor*,' came the shrill pipe of old Maria's voice from the little farmyard behind us, 'do not let the sheep eat those *Maria* weeds. They give the sheep terrible wind . . . AND A SHEEP CANNOT FART!'

This had truly been a morning of much learning.

– FIVE –

A DOG CALLED DOG

'HEY! HOWDY, *AMIGOS!* HOW'S YOUR BE-OO-TIFUL CHRISTMAS EVE? YEAH, THIS IS THE LASH COMIN' ATCHA ON *RADIO ONE – O – SIX – POINT – ONE FM* FROM THE STOODIOS OF *ANTENNA TRES* RIGHT HERE IN THE MIDDLE OF GOOD-OL' DOWNTOWN PALMA. YESSIREE – A GOOD, GOOD MORNIN' TO YA FROM SPAIN'S FIRST AND FOREMOST ENGLISH LANGUAGE RADIO STATION, BROADCASTIN' FROM THE SUNSHINE CAPITAL OF THE WORLD – BE-OO-TIFUL PALMA DE MALLORCA-A-A-A! HEY! YOU BETTER –'

Ellie popped her head round the door. 'Look, I'm sorry, but I just had to turn that racket off. I'm trying to get through to the airport to find out if the boys' flight is going to be on time this morning.'

'Oh, I'm sorry, I didn't realise. It's just that the "Oldies-but-Goodies Show" is coming on the radio after the news and . . . Hell's bells! What's that awful smell?'

'That? Oh, that's the dog food that Señora Ferrer gave me the other day. I took it out of the fridge a few minutes ago and stuck it all in an old pot to boil up for her precious Robin and Marian.'

'But what in blaze's name is it? I mean, it stinks like a badger's arse.'

'Well, there's a kilo of broken rice. That's what they use here to bulk out the dog food instead of biscuits and meal, and . . . you know . . . that sort of stuff.'

'Ellie, that stench isn't coming from rice, broken or otherwise, unless it's been lying in a midden for a couple of years.'

'No . . . well, the smell . . . the unusual smell will be the other ingredients that Francisca gave me. They're all OK – quite fresh and everything. She said they use these, ehm, *ingredients* all the time for pet food here.'

'OK, Ellie, just tell me what these other *ingredients* are, OK? Second thoughts, don't bother. I'll go and look for myself.'

The nauseating odour intensified as I entered the kitchen. Holding my nose with one hand, I gingerly lifted the pot lid and immediately dropped it on the floor. I hadn't seen a live hen looking at me from inside a boiling stew pot before.

'For Christ's sake, Ellie! What the hell have you got cooking here? There's a hen moving about in there. No two! Jeez! There's three of them!'

'Don't worry about it. It's all right. It's not a hen or hens. No, no, it's just . . . well, it's only hens' heads and feet . . . sort of swimming about in the rice, that's all.'

'Only hens' heads and feet! Ellie, you have got to be out of your mind. You've got a witches' brew going here that wouldn't be out of place in *Macbeth*, and the hum would be enough to fell an elephant at a hundred yards. And – and you say it's *only* heads and feet!'

'Well, don't blame me. That's what Señora Ferrer gave me – in that bundle, remember? I didn't even look at the contents at the time. Just stuck the bundle in the fridge and took it out this morning and –'

'But look!' I grabbed her arm and pulled her over to the bubbling cauldron. 'The bloody heads have still got all their feathers on . . . and combs, wattles, beaks, eyes, everything. And the feet – I mean, look at that foot going round just now. It's still got a plastic ring on its leg! You can't be serious about cooking up all that stuff in the house.'

Ellie looked at the floor and shook her head dejectedly. 'I know, I know. I admit that it maybe looks and smells a wee bit, well, disgusting, but that's what they use for dog food here. And anyway, what could I do? I could hardly take it back and tell her that I didn't want it, could I?'

'That's *exactly* what you should have done.'

'No. I'm sorry, but I just can't do that,' Ellie said assertively.

'And why not, may I ask?'

'Because, well, because Señora Ferrer would be insulted, that's why. To be honest, I think I was maybe a bit rude to her when I told her that her dogs weren't going to be allowed to sleep in here any more, and – well, she did give us the dog food because she genuinely felt it was her responsibility. She didn't want us to be out of pocket. Let's face it – she wasn't to know that we wouldn't like the smell of it cooking.'

'Wouldn't like the smell of it cooking? But, Ellie, that is precisely why the old farmers back home used to have a boiler in an outhouse for boiling pigswill. Because nobody would be daft enough to stink up the whole farmhouse by cooking all that garbage in the kitchen.'

'Fair enough, Mr Know-it-all, but we don't have an outhouse with a boiler here, so I *had* to do it in the kitchen. Right?'

'Right, Ellie, OK. I can see that you did it in good faith. But that's the first time, and it'll be the last. It was bad enough having her dogs crapping all over the kitchen floor every night, but even that didn't smell as bad as this.'

'Granted. And please don't imagine for one moment that I like it any more than you do. But I think you're forgetting one very important thing.' Ellie looked out of the window towards the Ferrer's *casita*. 'They're over there for the Christmas holiday, and you can bet that Francisca will be across here before long with another bundle of the same food for her pets.'

'OK. No problem. There's only one way to put a stop to that, isn't there?'

'Peter, you can't,' retorted Ellie with a look of astonishment. 'You mustn't offend her. You said it yourself – "When in Rome" and all that? Well, their way is to feed that stuff to their dogs and cats, and if you give them the impression that you don't approve, that you don't even appreciate –'

'I know, Ellie. Say no more. Your sentiments are very commendable, but there are times when you have to make a stand – times when you must be absolutely firm, even if it does cause offence. And this is one of them.'

'But you can't simply go and blatantly hurt their feelings.'

'Yes I can, and I will. First of all we had to put up with her dogs' mess, now she expects us to suffer the house being polluted by the pong of their revolting food cooking. It's just not on, Ellie. Just not on at all. You go and phone the airport. I'm off to see the Ferrers and give them a piece of my mind.'

The Christmas chaos at Palma Airport was true to form. French air-traffic controllers were having one of their normal peak-season huffs, so the vast arrival and departure halls were chock-a-block with the innocent victims – thousands of ex-holiday-making families, with whining kids and their distraught parents huddled together round their luggage like helpless refugees. *Bon voyage*, as they say in France.

Ellie and I cursed the entire French nation and headed for the bar in the National Departures Lounge, officially intended for the comfort of passengers on internal flights only, but the only bolt hole in the entire airport where anyone else lucky enough to know about it could escape the milling throngs of package-tour travellers. We settled down at a quiet corner table with coffees and *ensaimadas*.

'So how did your confrontation with the Ferrers go?' asked Ellie. 'You still haven't told me how they reacted to your low opinion of their "revolting" pet food.'

'No, well, there isn't much to tell really. As I said, it's merely a matter of being firm. Sensible people will always react in a mature way to that approach – as long as you're being reasonable and fair, of course.'

'Oh yes?'

'Oh yes indeed. Have no fear – you won't be getting any more hen heads and feet or any other dodgy poultry particles from Señora Ferrer. From now on she'll only be providing the broken rice.'

'Only the broken rice! But who's supposed to buy the rest of the food for her little menagerie, if you don't mind me asking?'

'Oh look! There's Jock Burns – the very man I wanted to see.'

'Just one second. Does Francisca Ferrer actually expect *us* to buy –'

'Hey, Jock! Over here! Got a minute, have you?'

Jock Burns was originally from our part of Scotland, but he'd lived in Mallorca for years, leading the quiet, respectable life of a teacher in one of the local international schools by day, and subsidising his hobby of eating, drinking and generally indulging in all the good times the island had to offer by working as an entertainer in tourist hotels at night. In more ways than one, Jock was a larger than life character. But his heart was in the right place, and he'd gone out of his way to help and advise us on many occasions.

I touched Ellie's hand. 'Sorry to interrupt you, but our Charlie has to start school out here in ten days time, don't forget, and there are one or two things on the subject that I need to check with Jock.'

Jock's entrance into the bar was characteristically over-the-top.

'Right on, Pedro!' he yelled, sweeping towards us between the tables with his loud yellow-and-red checked jacket draped movie-producer style over his shoulders, and calling out greetings in a contrived mid-Atlantic drawl to totally unknown and patently bewildered customers en route. He stopped in front of Ellie and opened his arms.

'Aw gee, Ellie baby,' he bellowed. 'What a woman! Hey, if you ever decide to ditch this bum here, I'm your guy.'

Ellie cringed, so Jock turned to me and clapped his hands to my cheeks.

'Pedro, my main man!' he proclaimed, raising his voice just a few decibels more. 'How are things in Hollywood, ol' buddy? Say, did you get my message from Clint?'

The fact that I'd never even been in the same hemisphere as Hollywood was of no importance to Jock. He slumped dramatically into a seat beside me, dangled his arms over the back of his chair and casually looked about for any signs of admiration from the nearby clientele. There was none.

'Fuckin' no-style plebs!' he muttered, lapsing into his natural Scottish accent.

I cleared my throat. 'I was going to ring you, Jock. A couple of things I wanted to know before Charlie starts at your school.'

'Aye, nae bother, son. I'll take care of everything. Just you turn up with the wee man on the first day of the new semester to pay the fees. Oh aye, and I've managed to get you a nice deal there, by the way. Nice wee discount.' He winked a slow wink. 'Save you a wee drop of money – know what I mean?'

'Well, thanks a million, Jock,' I grinned, slapping his back. 'Come on – let me buy you a drink.'

'Nah, nah – no time, son. Thanks all the same, but no time. Got to meet the London plane, see. Got a mate coming in with a brand new keyboard for me. Latest model. Magic. Half price and no questions asked – know what I mean?' Leaning towards us, he raised his eyebrows and lowered his voice. 'No receipt either, of course, so I've got to make sure my pal in Spanish customs turns a blind eye.'

'And we'll have to make a move too,' fussed Ellie, glancing up anxiously at the arrivals screen. 'Look – their flight! It's landed!'

'Ye're no half excited, eh,' teased Jock. 'Been missing your wee laddies, then?'

'They're hardly wee laddies now,' I said as we stood up from the table. 'Sandy will be nineteen soon, and Charlie's twelve coming on thirteen.'

'Well, they're still wee laddies to me,' confessed Ellie. 'And yes, Jock, I have missed them, and I don't mind admitting that either.'

'Well, you've not long to wait now for your big reunion,' Jock laughed. 'Aye, what a grand scene of unbridled Scottish emotion that's likely to be.'

'Oh, I almost forgot,' I said, tapping Jock's elbow as we ploughed our way through the swarming masses towards International Arrivals. 'A few things I wanted to ask about the school routine. Firstly, what should Charlie wear, and secondly –'

'Jeans and sneakers is the answer to your first question, and as far as the rest goes, just calm down and it'll all sort itself out. This is Mallorca, son. So keep the heid. Just be *tranquilo* – know what I mean?'

With that, Jock shouted out a protracted Americanised greeting to some non-existent 'buddy' at the other side of the hall. Then, when he was satisfied that enough people were looking at him, he swaggered off in a billow of yellow-and-red checks.

Ellie and I slunk away as inconspicuously as possible into the mêlée at International Arrivals.

'Hi, Mum. Hi, Dad,' came the familiar voices from behind us. 'We thought you'd forgotten to come and meet us. We've been here for over three minutes.'

'Sandy! Charlie!' squealed Ellie, trying her best to throw her arms around both boys simultaneously, only to have her loving salutations fobbed off by raised elbows and mortified squirms. 'You've no idea how much I've missed you both.'

'Yeah, yeah,' said Sandy automatically, 'but I'm completely skint.'

'I'm skint too,' added his younger brother, clapping his empty pockets.

'Yeah, but you were skint before you even got on the plane, you little twerp. I told you to lay off the computer games at the airport, but –'

'Now, now, lads,' I interrupted, 'it's supposed to be the season of goodwill and all that, so –'

'Hey, guys! Hang loose. I say, hang loose there, partners.' It was Jock Burns again, still in mid-Atlantic mode, and now lugging a suspiciously long flight case. 'Say, are you young dudes with some kinda group of Scottish diplomats or somethin'?'

'Hi, Jock!' said the boys in unison, suddenly all smiles as they held out their hands to give Mallorca's oldest teenager 'some skin'.

'That's right, Jock,' Sandy beamed. 'We're here for a convention – in the bullring.'

'Right on! But better watch ya don't scare the shit outta the matadors, kids.' Jock was revelling in this long-awaited appreciation of his 'act'.

'Anyway, I'm pleased to see that you managed to get your new keyboard through customs,' I said, pointing to his flight case.

'Nae bother, son. Just a question of knowin' who to touch wi' the old dropsy. That's how you survive on this island, by the way,' Jock confided, reverting to the ethnic Scottish vernacular as soon as we had emerged through the automatic doors into the covering din of coaches roaring in and out of the passenger pick-up area. 'But never forget ye're a foreigner, and always keep a low profile.' He gave me a wink of almost masonic confidentiality. 'Know what I mean, son?'

'Right, I'll remember that, Jock. Oh, and listen — I hope we'll see Meg and you during the holiday period. We'd appreciate your company.'

'Don't worry, pal. We'll likely catch you for a wee feed and a few bevvies durin' the festerin' season. Nae bother.'

There were handshakes, hugs and season's greetings all round, then Jock exited in customary style, saluting flamboyantly and calling 'Have a nice day now' to perplexed strangers peering down at him from a line of parked coaches as he sauntered theatrically by.

The boys had never seen their new home before, and as we drove through the gateway at Ca's Mayoral, the afternoon sunlight shining through the fronds of the old palm tree at the corner of the yard was casting exotic shadow pictures along the whitewashed walls of the house. The valley was still and quiet, save for the spirited chirping of a robin who had recently taken up residence in the garden and was now asserting his territorial claim from the topmost leaf of a giant rubber plant. Our own little harbinger of Christmas, Ellie called him.

Sandy and Charlie closed the car doors and stood looking silently at the surrounding scenery. I recalled how awe-struck I had been the first time I gazed up at those majestic mountains and sensed the enveloping feeling of peace and timelessness. I gave the boys a few moments to take it all in, to savour their first experience of this idyllic place, and then I said, 'OK, lads – what do you think? Quite something, eh?'

Charlie frowned and looked upwards, sweeping his hand vaguely round his head. 'All those mountains. Can't be very good TV reception. And grass. I don't see any grass. So where do we play football?'

'That's typical. Always thinking about yourself,' chided his elder brother. 'Selfish little git.' Then he turned to me and smiled reassuringly. 'Don't worry, Dad. It's all super. Very spectacular. Just one thing bothers me, though. How do we work in those tiny fields? Seems to me that there isn't enough room even to turn a decent-sized tractor. And how do you drive the tractor under all those fruit trees with the low branches?'

'Yes, well, that's another story, Sandy,' I sighed.

Both boys stood and stared at me, their heads inclined to one side, waiting for acceptable answers to their respective queries. How could I break the news gently that there would be no TV at all until the shippers had eventually delivered our long-overdue belongings from Scotland, that there was no grass on Mallorcan fruit farms, and that there wasn't any 'decent-sized' tractor, nor was there going to be anything to fit that description?

Ellie came to my rescue with a cheery, 'Come on then, boys. Grab your suitcases and follow me into the house. You'll be dying to see your new bedrooms.'

'Yes – good idea, Ellie,' I chipped in quickly. 'You show the lads the ropes. I noticed old Pep working in his *finca* across the lane when we drove up, so I think I'll take the opportunity to go over and introduce myself. I have to ask him about bringing his sheep in to graze down our weeds, anyway.'

Pep was leaning cross-legged against one of the big wooden-spoked wheels of his mule cart when I walked into his little shambles of a farmyard. He had just finished hand-rolling a crude cigarette which he stuck in the corner of his mouth and kindled up in a perilous-looking flare of flames and sparks, rather like the ignition of a duff squib.

This was the first time I had seen Pep close up, and it immediately struck me that he was probably considerably older than his slim, wiry physique had suggested from a distance. A small black beret was pulled forward over his forehead to shade his eyes – little dark beads shining out of a lean face of tanned, leathery skin. When he noticed me, his features contorted into a deep-wrinkled smile, revealing a graveyard of randomly-spaced teeth which were stained a rich brown by the smoke of whatever dried substance he was wont to burn in his roll-ups. Their smell evoked memories of years gone by, reminding me of the fumes from an old dung midden which once caught fire by spontaneous combustion on my grandparents' farm when I was a kid.

Although Pep's boots were almost worn through and his old grey trousers had patched knees and were frayed at the bottom of the legs, his general appearance was given a strangely dashing touch by his American-style bomber jacket – no matter how scuffed the leather – and by the little red neckerchief which was tied rakishly in a neat knot at his throat. I fancied that there was probably more to old Pep than the average Mallorcan *campesino*.

'*Buenos,*' he droned in a deep bass voice, rendered hoarse and husky, I guessed, by a lifetime of serious smoke inhalation. '*Cómo va su vida?*'

What a splendid greeting, I thought. *Cómo va su vida?* How goes your life? I would have to remember that one.

We exchanged the usual pleasantries, and I went on to explain who I was and where I had come from, but I got the distinct impression that Pep already knew all the relevant details anyway. He continued to lean against his cart wheel as I spoke, his thumbs thrust into the belt of his trousers while he squinted at me through one eye, keeping the eye above his smouldering cigarette firmly shut. Despite his laid-back pose, I knew that Pep was weighing me up very carefully indeed.

After I had completed the narration of my potted curriculum vitae, there followed an interlude of uncomfortable silence during which Pep maintained his monocular study of my face. The tension was only relieved when he choked on his cigarette smoke and coughed a cascade of sparks down the front of his jerkin. But this didn't faze Pep. He merely croaked a chesty '*Va bé, va bé*' (the *mallorquín* equivalent of 'OK, fine'), re-crossed his legs and continued to scrutinise me – though now through both eyes.

Getting on the right side of this old codger was not going to be easy, I suspected. 'That's a good-looking dog you've got there,' I said with forced enthusiasm, nodding towards the motionless black shape lying in the sun outside the door of Pep's dilapidated cottage.

Pep's eyes lit up. I had obviously stumbled on the subject nearest to his heart – his animals. He looked admiringly at the huge, sleeping beast, and began to sing its praises. That dog, he could confidently state without fear of contradiction, was a perfect specimen of Mallorca's famed native breed, the *Ca de Bestiar* – an animal of unique intelligence, a *perro* of many qualities. *Hombre*, no better sheep dog could be found anywhere in the world. But even more amazing was the breed's hereditary alertness and its matchless bravery as a guard dog. *Sí*, it was fearless, the *Ca de Bestiar*. *Molt valent!* 'Just wait till you see the size of this one,' he said, scarcely able to contain his enthusiasm. *'Qué energía!'* He cleared his throat and shouted, 'Perro! Oy, Perro!'

The dog never moved a muscle.

'Perro – isn't that an unusual name for a *perro*?' I suggested, breaking the silence. 'You know – a dog called Dog?'

'Unusual? Certainly not! I have had many *perros*, and I have called them all Perro. Simpler that way. Unusual? *Mierda!* Perro!' he yelled again, plucking the cigarette from his mouth in order to facilitate the production of maximum volume from his kippered vocal chords. 'PE-E-RR-RR-O-O!' Nothing.

Pep bent down, picked up a sturdy stick and hurled it with all his might at the comatose dog. There was a hollow 'clonk' as the projectile hit Perro solidly on the top of his head. I feared for the animal's life, but the *valent* Perro only flicked one ear as if shaking off a fly, made a few contented clicking sounds with his tongue, and continued his *siesta*.

The *perro* was very young, Pep explained. Only eight months old – a pup – still under training. But I should not be fooled by its present appearance of lethargy. That was only because Perro had eaten three whole pigeons for lunch. 'Broke into the damned cage,' Pep mumbled almost inaudibly. *'Hijo de puta!'*

'He's certainly a magnificent specimen,' I said brightly, in an attempt to cover any embarrassment that Pep might be feeling because of Perro's total disregard of his master's voice.

But Pep's confidence was in no need of any boosting from me. 'Just remember what I say,' he advised sternly. 'Those *perros* are very suspicious of strangers. Very unpredictable, potentially lethal, the *Ca de Bestiar*. In the breeding, you understand. *Cuidado!'* I had been warned.

'Gracias, Señor Pep. I'll bear that very much in mind.'

I thought that it was then about time to guide the conversation towards the real reason for my visit, but I wasn't quick enough for old Pep. Oh no.

I would be wanting him to bring over his sheep to graze my fields, he conjectured without prompting. *'Va bé, va bé.'* That would be no *problema*. Some of his *ovejas* had just lambed so he was looking for clean grazings nearby, and he didn't want to walk them too far in any case, because there was bad weather on the way. *Sí, sí*, a pasture close to his *finca* would be *muy conveniente*.

'Muchas gracias, Pep. I'll leave the field gate open for you, then.'

There was just one tiny *problema*, he cautioned. The field nearest the *torrente* – the field with the well; I would have to plough that one without it being grazed first. Too much of the Maria weed. Very bad for the sheep. Gave them the bloat. He farted loudly to emphasise the point. *'Comprende?'*

I couldn't help smirking. This old boy was a real rough character, but there was something likeable about his style.

'You can afford to smile,' he said gruffly, obviously mistaking my expression for one of derision, 'but you would not think it so funny if you were a sheep. A gutful of Marias . . . VABOOM!' He threw his beret in the air. *'Una bomba!'*

I apologised for upsetting him. It had been quite unintentional, I stressed, also mentioning that Señora Bauzá had already warned me of the dangers of grazing the Maria weed. Her approach had been slightly different, of course, but the basic message had been the same.

Pep chose to turn a deaf ear. He spat the burnt-out stub of his cigarette onto the ground and fished in his inside pocket for the makings of a fresh one.

'Cigarrillo?' he asked, offering me a flat tin box containing a packet of cigarette papers and an evil black mash of shredded vegetation which looked as if it had been dredged from the bottom of a compost heap.

'No . . . no *gracias*,' I insisted, pointing out that I had given up smoking years ago.

Pep was clearly perplexed. So what? Even if I had given up years ago, he speculated with an exaggerated shrug, how could I possibly think of refusing a smoke of this stuff?

He stuck his forefinger and thumb into the tin, teased out a wad of the noxious shag, and stuffed it under my nose. 'Sniff that, *hombre*. Sniff that.'

I obeyed, and I wasn't lying when I told him that it smelt good; sweet, strong, pungent even – but good. How mysterious, therefore, I mused, the metamorphosis into such an acrid stench when wrapped in paper and set alight. There was little doubt in my mind that, of necessity, Pep would chew a handful of raw garlic cloves every morning just to freshen up his mouth. So no, I still wouldn't have a smoke, thanks all the same.

Visibly miffed by my rejection of his generous offer, Pep readopted his resting position against the cart wheel and proceeded to fashion another *cigarrillo* for himself.

Had I any idea what I was turning down, he asked without raising his eyes from the delicate rolling process? Was I not aware from the aroma that this was genuine Havana *tabaco*?

He licked the paper, ran his finger carefully along the join, screwed one end of the lumpy cylinder into a wick, moistened the other end liberally with his tongue to stop the paper sticking to his lips, then wedged the virgin *cigarrillo* in the corner of his mouth.

Well then, what did I have to say about that? Did I not know where Havana was?

'Why, yes – Cuba,' I answered sheepishly. 'I know that the best cigars come from Hav–'

'*Sí*, and I have been there,' he interrupted, striking a match with his thumb nail and staring at me unblinkingly through the minor pyrotechnics display which took place less than three inches from his face after he lit the *cigarrillo*'s touchpaper. As he drew in the thick smoke, a strangled, almost imperceptible snort issued from somewhere at the back of his nose, and although he probably believed that he had successfully disguised a desperate urge to cough, his instantly-bloodshot eyes and the veins standing out like blue serpents on his forehead were a dead give-away. Pep's lungs thought his genuine Havana *tabaco* was made from old cabbage leaves.

'At one time,' he spluttered in a strained, high-pitched voice, 'many Andratx men went to Cuba to fish for the sponge. Just went for a few years to make money.' He took a deep gulp of fresh air and swallowed hard. 'But some never came back to Mallorca. Left their wives here and never came back. *Me entiende?*'

'Yes, I follow. But why didn't they come back? Were they prevented from returning when Castro came to power in Cuba?'

'Castro? Castro? Nothing to do with Castro. *Hombre*, they just did not want to come back to their wives.' Pep took the cigarette from his mouth and looked at it dreamily. 'But I came back. Ah *sí*. I came back after three years to marry my *novia* here in Andratx.' He sighed deeply. 'And what have I got to show for it now? My *tabaco* . . . only my *tabaco*.'

'Oh, I'm very sorry to hear that. I hadn't realised that you'd lost your wife, so please –'

'This is genuine *tabaco de Havana*,' butted in Pep. 'It is free and it gives me continual satisfaction. Show me a wife with those attributes. *Sí*, show me a wife like that.' He stuck the cigarette back in the corner of his mouth and fixed me once more in a penetrating one-eyed stare.

'Well, if you put it that way. But Havana tobacco – how can it be free? I don't understand how –'

'*Caramba!*' It was free, he told me impatiently, because he grew it himself. That was why. He had smuggled some plants into Mallorca when he returned from Cuba all those years ago. Very risky. Franco would have flung him in jail if he had found out. Anyway, he had planted them all, but despite all his careful and tireless nurturing, only two of the plants had survived. But from those two, he had developed the entire plantation of fifty tobacco plants which now flourished behind his *casa*. *Gracias a Dios*.

'Fifty plants, eh? That'll keep you puffing all right. But how about curing the tobacco? How do you do that?' I asked, politely feigning interest.

A guttural chortle rumbled from the depths of Pep's chest. He waved his forefinger slowly in front of my face. Surely I did not expect him to reveal such valuable information, he scoffed. That was a secret which he had learned from an old Cuban *negro* just before he died. Creeping consumption, they said. Too bad. Never mind, many men had offered huge bribes for the formula, but Pep would never divulge it. Not for any price. The secret would go to the grave with him.

Yes, and that day might not be too far off, I thought as I watched Pep venting another blue cloud of the putrid reek through his nostrils. Keeping the secret formula to himself might well constitute a selfless and valuable contribution to the health of countless future generations of Mallorcans.

Pep leaned towards me and lowered his voice to a gravelly whisper. 'There are men here in Mallorca who would like very much to steal a few of my plants in order to start their own plantations of *tabaco de Havana*. *Hombre*, they have tried before, many times. That is why I always keep an attack dog about the place.' He looked admiringly at Perro, who slept blissfully on, his legs twitching as he chased the dreaded tobacco bandits through dreamland.

A sudden chill breeze blew down the valley, singing through the pines on the high mountain slopes and whipping up a cluster of mini whirlwinds which zigzagged over the little farmyard in tiny twists of dust and straw. It only lasted a few seconds, then all was still once more.

Pep looked towards the northern sky and announced sagely, '*Sí, ah sí* – a change in the weather is coming. It will be cold this night. *Sí, mal tiempo viene.*' He puffed a gravity-defying length of ash from his *cigarrillo* and glanced dispassionately towards Ca's Mayoral. 'And you have no firewood, have you?'

'Ehm, well, there may be a few bits of wood,' I said, totally taken aback. 'Señor Ferrer did leave some for us, I think.'

'Hrumph! He must have made a mistake then. I saw Ferrer wheeling barrowloads of logs from your place to his *casita* the day before you came. You will get nothing from him. *Nada*. Talk about being mean? *Coño*, that Ferrer would not even piss on you if you were on fire.'

So, no matter how highly regarded the Ferrers may have been in the Andratx area generally, it was becoming increasingly obvious that their stock with their immediate neighbours in the valley could hardly have been lower.

'I – I'd better order some logs right away, then,' I said impassively, not wishing to take sides in Pep's apparent feud with Tomàs Ferrer.

'Order logs?' Pep shook his head. '*No es posible*. The sawmill is closed for *las vacaciones*. You will get no firewood from there until next week.' The end of his *cigarrillo* crackled angrily and glowed a bright warning red as he sucked in another chestful of genuine Havana fumes. '*Hombre*, you are going to have a cold, cold Christmas.'

What could I say to that? I was beginning to feel like the original *loco extranjero* again.

'*Bueno,*' said Pep, nodding in the direction of the setting sun. 'I must go inside now and light the fire before the temperature starts to drop. Must keep the house warm – especially at Christmas, no?' He didn't even bother to say goodbye.

What a lousy old Mallorcan Scrooge, I thought to myself as I watched him walk away. It was bad enough to point out to me that I was out of logs, but then to rub my nose in it like that . . .

Pep's shoulders started to shake, and I heard what sounded suspiciously like a chuckle – mucus-wrapped maybe, but a chuckle sure enough. Yes, the nasty old sod was having a good laugh, and at my misfortune, too!

He turned round, his eyes streaming. 'Ha! Your face!' he guffawed, one hand clutching his heart, the other pointing at me. 'What a picture! *Madre mía!*'

I was speechless. I could only stand there feeling and looking silly until Pep had laughed himself out. I didn't know enough Spanish obscenities to call him the things I wanted to.

'Hey, did you really believe that I could be such a bastard?' he eventually wheezed, wiping his eyes with his sleeve and hacking freely. 'Do not forget, *amigo* – my name is not Tomàs Ferrer, eh!' He walked up to me and grabbed me by the shoulders, wheeling me round to face the gate. 'Look over there by the lane. I have put a pile of logs there for you – enough to last you until the sawmill opens again. You have two sons, I see – so you had better send them over to carry the wood into your house before it gets dark. Believe me, it is going to be cold this night – cold enough to freeze the tits off a wooden saint!'

I was quite touched, and almost lost for words. I tried to thank him for this unexpected act of premeditated kindness, but he would have none of it.

'*Hombre*, what are neighbours for?' he grinned. 'I may not be a rich man, but what little I have is yours – if ever you are in need. I have more than enough wood on this *finca* for my own requirements, so *no problemas. Va bé.*'

He sauntered back towards his cottage, turned and threw me a John Wayne cavalry salute. '*Adiós, amigo.* May your life always go well and your hens always lay double-yolkers. *Adiós, y Feliz Navidad!*'

'Merry Christmas to you too, Pep. And thank you sincerely.'

Feeling both humbled and cheered by this spontaneous act of generosity, I headed for home and was almost out of Pep's farmyard when his voice rang out again:

'PE-E-RR-O-O-O!' he yelled frantically. 'PE-E-RR-O-O-O!'

Suddenly, I heard the thud of galloping feet behind me, and I spun round just as the massive, black dog launched itself off the ground and came hurtling at me, head high. Everything seemed to slip into slow motion. The dog's huge mouth was gaping open, saliva dripping from his tongue, and I was staring into powerful jaws armed with rows of deadly white teeth – aimed directly at my throat.

The wind was knocked out of me as the beast's heavy front paws hammered into my chest, and I was propelled backwards over the heap of logs, landing helpless on my back with my feet in the air. The dog pounced on me, pinning my shoulders to the ground, its fetid breath hot on my face.

I could hear Pep hollering wildly somewhere in the distance, but it was no use. The animal was out of control. It had me where it wanted, and all I could do was pray that the end would be quick and, above all else, relatively painless. I felt the crazed beast's wet nose at my ear, and I froze terror-stricken as I waited for its teeth to sink into my jugular.

Time stood still for seconds that seemed like hours. But nothing happened. I could feel the huge paws pressing relentlessly into my shoulders, I could hear the feverish panting. But nothing had happened. I wasn't being savaged. I was still alive.

I cautiously opened one eye, at which the dog let out a booming 'WOOF!' But instead of tearing me to shreds, he just looked down at my face, his great head cocked inquisitively to one side and his tongue dangling stupidly out of the side of his mouth.

'Perro?' I said hesitantly.

'WOOF!'

'OK, Perro. Just take it easy – there's a good boy. Just you let me up now. That's a good boy,' I coaxed in a tone of exaggerated calm.

But Perro could contain himself no longer – or perhaps, I wondered too late, he simply didn't understand any English. He lunged forward and started to maul my face . . . with his tongue. He licked and slobbered over everything that wasn't covered – my eyes, my hair, my nose, my mouth, my neck. There could be no more doubt, this attack dog of Pep's was nothing but a big pussycat.

As the panic drained from me, I started to laugh and giggle. Perro was licking me so much I could hardly breathe, and it tickled.

'OK, Perro, OK,' I panted, rubbing his floppy ears playfully. 'You'll lick me to death, big, soft lump that you are.'

It took but a second to realise that rubbing his ears had been a fatal mistake. I felt a warm sensation on my legs, which took me back to the day we first discovered this valley, the day when I shook hands with Señora Ferrer's hose-pipe. But it was a warm sensation this time – warm but just as wet. Dear God, no, I prayed, it couldn't possibly be – could it? I looked along the underside of Perro's straddling body, and my worst suspicions were confirmed. This lanky, overgrown idiot of a pup was peeing with excitement – all over my flailing legs.

A hefty chunk of timber whizzed past my face and bounced off Perro's skull with a resounding wallop. Then Pep's dusty boot came into view, moving deadly fast and directed upwards at Perro's protruding and unsuspecting scrotum.

The big dog yapped a strangulated, adolescent yelp as the kick landed bang on target. He peed a final shocked squirt on me, then lolloped off three-legged in the direction of his house. He would have a more practical use for his tongue now.

'I am truly sorry, *amigo*,' gasped the distressed Pep, helping me to my feet. 'I got him off you just in time – before he went in for the kill. It is *infortunado* that my dog attacked you, but he was only doing what he was bred for, you understand.' He laid a friendly hand on my shoulder. 'He could not help it, *amigo*. It is in his genes.'

'I'm not too bothered about what's in *his* genes. It's what the goofy mutt has done on *my* jeans that bugs me,' I grumbled in English.

'*Correcto,*' agreed Pep, not having understood one word of what I'd said. 'He is a born champion. Unpredictable, lethal, but a champion. *El Ca de Bestiar . . . qué perro tan magnífico!*'

Ellie was chopping up vegetables in the kitchen when I staggered back in.

'What happened to you?' she asked. 'You look as if you've been rolling about in a ditch.'

'Don't talk about it.'

'And I see you've wet your trousers again. Getting to be a bit of a habit of yours here, isn't it?'

'Just forget it, Ellie, OK?'

'And you stink too. Phew! That's strong. You smell like a stable drain.'

'All right, Ellie. You've made your point. Just drop the subject now. OK?'

'Well, it could be a sign. Maybe it's time you tempered your drinking habits. You know what I mean – when your urine starts to smell like a horse's, there must be something wrong. It's just not . . .' (She paused to cross-cut an abnormally large carrot in two) '. . . normal.'

'There's nothing wrong with my drinking habits, and it's not *my* urine anyway.'

Ellie stopped chopping and looked me up and down. 'Are you actually suggesting that a horse lifted its leg and piddled all down your jeans?'

'Something like that. But you wouldn't believe me, even if I told you. Just forget it. I'm off to have a shower.'

At that, the kitchen door burst open and the boys bundled in.

'I still think Sandy's room is better than mine,' griped Charlie. 'Oh, and he's got a bigger bed too.'

'Stop moaning, squirt,' said his big brother, clipping him on the ear. 'You're lucky to have a bed at all. You'll only pee it as usual.' He sniffed the air and screwed up his face. 'Jeez, Mum! Talking about pee . . . who let the horse in?'

'Oh yeah! Out of order!' proclaimed Charlie, holding his nose. Then he noticed my wet jeans, and his eyes opened wide in a startled look that harboured undertones of confused admiration. 'Aw, Dad – it's you. You've pissed your pants!' He mimed vomiting. 'Disgusting!'

'That'll do, Charlie,' I snapped. 'If you're looking for a belt on the other ear, you're going the right way about getting it.'

I looked down at my fouled jeans and added ruefully, 'Anyway, it wasn't me. It was Pep's dog.'

There was a silence in the kitchen as they all stared blankly at me.

Sandy eventually broke the ice. 'Never mind, Dad. I believe you, even if Mum and Charlie don't. But just one thing –'

'Yes?'

'Just what possessed you to allow Pep's dog to wear your jeans in the first place?'

The kitchen echoed with hysterical laughter.

'Right. Enjoy yourselves,' I said blandly, putting on an act of bored indifference. 'I don't mind being the butt of all your mirth. My shoulders are broad enough. But once you've pulled yourselves together, you two can go over and carry in the logs which you'll find at Pep's gate. Now, I'm going to take a shower.'

'Ah-ehm, before you go, dear – I have a message for you from Señora Ferrer,' Ellie said offhandedly.

'Oh yes?'

'Yes. She brought this for you while you were over at Pep's place.' Ellie handed me a boxed bottle of 'Gran Duque d'Alba' brandy.

'Well, well! Expensive stuff!'

'Uh-huh. And she brought this for the rest of us.' She clattered a tin of Francisca's dreaded homemade almond cake onto the table. 'So you can see who's the most popular person in this family as far as the Ferrers are concerned.'

Charlie sniggered expectantly while Sandy looked down at his feet and tried to suppress a grin.

I chose to remain silent, on the grounds that I sensed another mickey-take coming up.

'And her message for you is this . . .' Ellie stalled and eyed me closely to ensure that I was suffering maximum apprehension. 'Her message is that she and Tomàs were totally surprised and absolutely delighted by the bottle of Scotch and the wonderful fruit cake which you handed in to them this morning.'

'Oh that? I – it was only – well, I had to . . .'

'And there's more. She also thanks you from the bottom of her heart for the kind offer.'

'The eh . . . kind offer?'

'Uh-huh. The kind offer you made to save her the expense of buying hens' heads and feet for pet food.'

'Oh, that offer?'

'Mm-hmm. That offer. She says we're very fortunate to know such a friendly English butcher in Palma Nova, and we're quite right not to displease him when he has insisted on giving us unlimited free scraps of meat to feed her dogs and cats. *Muy generoso*, I think she said.'

'Yes, well it seemed like a good way of –'

'Yes, but we don't know an English butcher – friendly or otherwise – in Palma Nova or anywhere else on this island, do we, dear?' Ellie was tightening the screw.

'No – no, but I'm sure we can find one,' I mumbled, duly crestfallen.

'Yes, of course we can, pet,' cooed Ellie with more than a hint of sarcasm. 'You've done very well and I shouldn't be pulling your leg so cruelly.' She patted me on the head, then kissed my cheek.

Charlie faked another vomit.

'Cut that out, Charlie,' I warned.

'That's right,' said Ellie, just a bit too resolutely, 'you should be proud of your father. In one bold move, he's restored our friendly relationship with the Ferrers and he's spared us the sickening prospect of boiling up bits of dead hens for all time to come. Let this be a lesson to you both. Never forget your father's words . . . "There are times when you have to make a stand – times when you must be absolutely firm."'

'And times when you have to give the enemy bottles of whisky and fruit cake?' suggested Sandy.

'That's known as diplomacy, son – diplomacy,' I grunted.

'But Mum said she bought that fruit cake for us,' objected Charlie. 'It was supposed to be our family Christmas cake, and it's my number one favourite too. Now this old Ferrer bint's got her choppers into it and we're lumbered with her crappy almond stuff. That's not diplomacy. That's nuts!'

Anticipating physical violence, Sandy hauled his brother smartly through the door and off to collect the firewood.

'Pay no attention,' said Ellie soothingly. 'There's plenty more fruit cake where that came from. And we can always use the almond cake to help feed the dogs and cats again – until you find your mythical philanthropic butcher, that is.'

I went and took a shower.

Pep's forecast was correct. The temperature did drop dramatically that evening, and we were very glad that he'd thoughtfully provided us with logs.

The inglenook in the kitchen was really a little room within a room – a cosy square enclosure built around the wide corner hearth and formed by high-backed stone benches, which served both as comfy, cushioned seats facing the heat and as an effective barrier against the chill draughts which haunt old, stone-floored houses in winter.

We had kindled the fire in the traditional Mallorcan way, lighting a little heap of dry almond husks and adding increasingly large twigs and sticks until enough flames had been generated to ignite one of Pep's huge, olive wood logs.

The open hearth was topped by a bell-shaped chimney of whitewashed plaster, and fixed to the walls on either side were rustic wooden shelves holding an array of earthenware pots, jugs and plates between which Ellie had placed our Christmas cards and some sprigs of holly. A bushy pine branch which we had dragged down from the mountains a few days earlier stood in a tub by the door, transformed into an improvised Christmas tree by the addition of a string of fairy lights and a scatter of tinsel.

'It's so cold, I think it would be nice to have our Christmas Eve dinner here round the fire,' said Ellie, hanging a big iron pot on a metal bracket which she swung out over the flames. 'I've got some real old-fashioned Mallorcan treats, and you can all help yourselves.'

She dipped into a basket and pulled out a mouth-watering selection of pastries that she placed carefully on a grill over a mound of glowing charcoal at the side of the fire. 'I bought these freshly baked today from the Panaderia Tony in Andratx, so all we need to do is heat them up.'

There were *Empanadas mallorquinas*, delicious little individual lamb, pork and *sobrasada* sausage pies. There was *Coca mallorquina*, the old Mallorcan version of pizza, made from the baker's surplus *ensaimada* dough and baked in the bread oven till the lightly browned pastry is crisp and flaky. The hanging pot contained Ellie's first attempt at *Potaje de Lentejas*, a warming wintertime pottage of lentils, vegetables, *sobrasada* and the spicy Mallorcan black pudding, *butifarrón*. The recipe was a tribute to the simple ingenuity of Mallorcan country cookery, and had been given to Ellie by old Maria Bauzá, along with a little jar of her *fonoi marí* which, she insisted, was also an essential accompaniment to this particular traditional dish.

The *Potaje* was a huge success, and the boys ladled it into their *greixoneras* with undisguised gusto while the fire hissed and crackled in front of us, sparks spitting from Pep's yule log and dancing up the chimney in the rising heat from the roaring blaze. Ellie dipped into her basket again and produced some chestnuts, spreading them out to roast on the grill which had already been stripped bare of the savoury pastries by our ravenous progenies, while I popped the cork of a bottle of *Herederos de Ribas Tinto*, a wine from the little inland town of Consell.

After the boys retired full-bellied to bed, Ellie and I sat on for a while in the inglenook, watching the brilliant colours of the dying flames lapping gently now over the remains of the charred firewood, occasional down-draughts of cold air from outside blowing little puffs of aromatic smoke into the room and spinning the wood ash on the hearth into tiny eddies of grey-white powder.

'It's good to have Sandy and Charlie here,' Ellie said.

'Happy families, eh?'

'Yes, and I've missed them − even with all their silly squabbles. After all, home is where your family is.'

'True − and it could have been worse.'

'How do you mean?'

'Just think about it. If I'd discovered *Hierbas*, snails and wormy orange juice years ago, we might have had half a dozen of the buggers scoffing our mince pies tonight.'

'Time for bed,' yawned Ellie. 'It's been a long day.'

'Yeah, and it's getting chilly now that the heat's out of the fire. As old Pep said, it'll be cold enough to freeze the −'

Ellie stopped at the top of the stairs and gripped my arm. 'Sh-sh! Listen − there's somebody out there!'

'It'll only be one of Francisca's cats on the prowl. Forget it.'

'No − there it goes again. It's a voice − a human voice − all gruff and . . . creepy!'

'You're imagining things. Come on − it's freezing cold standing here. Let's get to bed.'

'No – there's no way I'm going to bed until we – you – find out what's out there. Go on – go out on the balcony and take a look.'

'OK, if it'll make you happy. But I'm telling you, you're imagining things,' I said hopefully.

I cautiously opened the French doors and poked my head outside into the darkness.

'So – can you see anything?' Ellie hissed. 'Is there somebody out there?'

I nodded and beckoned her towards me, putting my finger to my lips. 'Come outside,' I whispered. 'It's all right, but be as quiet as you can.'

Ellie took my hand, and I could feel her pulse racing as she stepped hesitantly onto the balcony.

'Listen carefully,' I murmured. 'Can you hear it?'

She shivered and clung tightly to my arm. 'Yes . . . it's a voice . . . a man's voice – all deep and spooky and moaning and groaning down there. It's scary. I'm off!'

'Just stay put and keep listening. What else do you hear?'

'Bells. I can hear bells,' she said softly, relaxing her grip on my arm and moving over to the balustrade. 'Oh . . . it's sheep. I can just see them now – grazing in the little field in front of the house. And look, some of them have got lambs.'

'Mmm. It's old Pep and his little flock,' I whispered. 'But just stand quietly for a moment and look and listen. I think there's something very special about this.'

The night was perfectly still, the starlight etching the high mountain ridges into dark, rugged silhouettes against a clear and cloudless sky, the valley hushed and sleeping. At the far corner of the field, we could just make out the shadowy figure of Pep leaning against the twisted trunk of an ancient olive tree, his gaunt frame draped in a long blanket to keep out the cold. He was talking gently to his sheep – watching them, listening to the sound of their bells and tenderly calling back any that might have strayed too far from the flock, just as shepherds had done for thousands of years.

Ellie and I said nothing, but we both knew that our thoughts were being drawn inexorably to the story of just such a night, a silent night, a long, long time ago.

The chimes of midnight drifted up the valley from a distant village church, and a newborn lamb cried to its mother in the field below.

It was Christmas Day.

– SIX –

NIGHTMARES FOR CHRISTMAS

The weather on Christmas morning lived up to old Pep's forecast. According to young Charlie, the scene was more like some dreich dump in the Scottish Highlands than the sun-drenched paradise valley he had been looking forward to. This was the pits, he judged, and not even a telly to look at.

Bad weather didn't last for long in Mallorca, I assured him, looking out of the open kitchen door at the threatening rain clouds accumulating over the mountains. The sun would soon be shining again, then he could really start to enjoy the open-air Mediterranean life.

'Just think of the fun we're going to have on this little farm,' I enthused. 'There's all the orange-picking to do – you can earn a few pesetas at that – and once all the weeds have been grazed down and ploughed under, we can fix up a bit of field for playing football. It'll be a dirt surface, of course, but that's normal here. So come on, cheer up. Everything's going to be fantastic.'

'All the lads back home will be watching TV right now,' mumbled Charlie, clearly unimpressed by the promise of orange-picking opportunities. 'There's always loads of good programmes on TV at Christmas.'

'Yes, but you'll soon forget about all that once the weather gets back to normal. It's just a completely different lifestyle here. You'll see . . . going to school in jeans and sneakers, swimming all summer long –'

'And football pitches with no grass. Don't fancy that much.'

'Don't worry, you'll soon get used to it,' I smiled.

'Don't think so,' frowned Charlie. 'I'd have to get used to catching gangrene too. You can catch that playing on dirt pitches. Something to do with sheep droppings. You end up having to get your legs cut off.'

'We can only hope that you get gangrene of the tongue as well,' muttered Sandy as he wandered into the kitchen rubbing the sleep from his eyes. 'Then some of us might get peace to go to sleep at night instead of listening to all your moans and groans through the wall. You don't know when you're well off, that's your problem. Never satisfied.' He glanced out of the window and shuddered, raising his shoulders and rubbing his hands together. 'Bit grim out there, isn't it, Dad? Not quite your sun-drenched paradise valley, eh?'

'*NARANJAS! HOLA! NARANJA-A-AS!*'

I recognised the call of old Rafael. He had arrived at the gate just in time to save me listening to any more disparaging comments from the boys about the disappointing turn in the weather. What the hell did they expect, I asked myself: non-stop sunshine for three hundred and sixty-five days a year?

'*Feliz Navidad, amigo. Cómo le va?*' beamed Rafael as I opened the gate for him. For once, he wasn't accompanied by a mob of goats or grandchildren. He was in a hurry today, he pointed out, so he had come on his motor scooter instead. And where had I been anyway? He had been up to buy oranges yesterday, but nobody was around. *Hombre*, I could never expect to make money in the fruit business if I wasn't around when the paying customers called. '*Madre de Dios!*'

I was standing downwind of Rafael while he heaved and kicked at his ancient scooter until he was satisfied that it was resting securely on its wobbly, one-legged stand. The smell of goats was overpowering, and the thought occurred to me that perhaps Rafael actually ran this machine on some sort of secret fuel distilled from goat urine in his own back-yard refinery.

'Sorry I missed you yesterday, Rafael,' I said, moving discreetly upwind, 'but we had to go to the *aeropuerto* to pick up our two sons. I'll send them down to the orchard to help you pick your oranges. It'll save you some time if you're in a hurry.'

That would be good, he said. He would appreciate some help today, because he had a lot to prepare before the family came for Christmas lunch. Only twelve of them this year, but still a lot to prepare. And anyway, there was some really bad weather brewing. 'Just look up there, *amigo*.'

The clouds had begun to billow into swelling masses of foreboding grey, obliterating the topmost mountain ridges under a ghostly mist which was already creeping down the high gullies, extending its cold fingers slowly but relentlessly through the pines to the valley below. High above, smaller, fragmented clouds were scudding across the darkening sky, heralding the arrival of the *Tramuntana* wind and its companion storm clouds that could already be seen gathering menacingly over the far northern mountains.

'Grab some secateurs and a couple of baskets, lads, and nip down to the orange groves to give old Rafael a hand,' I said, rushing back into the kitchen. 'He's a bit pushed for time.'

'But, Dad, we don't know anything about picking oranges,' objected Charlie.

'So who better to learn from than an old Spanish orange-picker?' I retorted.

'Yeah, but the language, Dad. I mean, does this old guy speak any English?' Sandy asked apprehensively.

'Not one single word, so it'll be a good chance for you to practise your Spanish. You've got to start sometime, and there's no time like the present, so go for it.'

'But our Christmas presents – we haven't even opened our Christmas presents,' the boys wailed in unison.

Their protests fell on deaf ears and they sloped off reluctantly, newly united in their common lack of enthusiasm for the alien task in store, and for – as I heard Charlie grumble under his breath – the whole poxy Mallorcan scene.

'Just be patient with them, dear,' said Ellie, who was busy raking the fire and tidying up the inglenook after the previous evening's feast. 'It's all a big change for the boys. It'll take time for them to settle. Just be patient with them.' She smiled a patience-inducing smile and added, 'You know, I think you should re-kindle this fire while there's still some heat in the ashes. We want to be nice and cosy on Christmas Day, particularly when the sky looks so overcast and threatening.

We must make the most of the few home comforts we've got at present, mustn't we?'

She was right about that. In the absence of all our familiar bits and pieces – the pictures, the lamps, the ornaments, Charlie's beloved TV, all the little things that make a house a home and which were still in transit somewhere between Scotland and Spain – we really did have to create as much cheer as we could. And there could be nothing better than a roaring log fire at Christmas, apart perhaps, from a houseful of friends and relatives and well-wishing neighbours. But all of that was far behind us now. It was all fifteen hundred miles away, and it might just as well have been fifteen thousand miles, I suspected, as far as Charlie was concerned.

'Charlie will soon adjust,' said Ellie, reading my thoughts perfectly. 'Once he starts school, he'll make new friends and find new interests. It'll be fine, I'm sure. But Sandy . . . well, at his age, it could be more difficult.'

'Yes, I know what you mean,' I said, fanning the embers to ignite a little pile of almond shells. 'He had already started to establish a life of his own at home – a bunch of good mates he'd known all his life, his first car, even if it was a heap, and his plans to go to college next year. It must have been a difficult decision for Sandy to follow us out here.'

'Mm-hmm. I do still wonder if we made the move at the wrong time for Sandy,' replied Ellie pensively. 'But then, it *was* his own decision to join us, and knowing him, he'll be keeping his options well and truly open.'

'How do you mean?'

Ellie knelt down beside me on the hearth and began to puff encouragingly at the crackling almond husks with an old set of leaky bellows that the Ferrers had left behind. 'Well, it's not as if he's been sentenced to life out here, is it? It's not as if he's been banished to some penal colony for the rest of his days. If he doesn't like it or if he doesn't see a good future for himself here, I'm sure he's sensible enough and strong-willed enough to go back and pick up where he left off.'

'Yes, but I remember you and I going over and over this so many times when we used to sit and fantasise about coming to live in Spain, and putting pressure on Sandy to make a decision about staying behind or coming with us was always our greatest worry. I know that Sandy encouraged us more than anyone to make the break, but now that we're actually here and he's burned his boats by coming out too, I just wonder –'

'Look,' said Ellie, placing her hand on my arm, 'you're worrying yourself unduly. Don't forget Sandy's not a wee laddie any more.'

'No matter what you said to Jock at the airport?' I teased.

'No matter what I said. Sandy's a young man now and he's got a good head on his shoulders. He'll do what's right for himself, I'm sure.'

'You're right, Ellie, you're right,' I sighed. 'It's just that it's only now that I'm beginning to realise what a big change this is going to be for the boys. It doesn't matter so much for you and me, but coming to live here is going to affect the rest of their lives.'

'Hey, snap out of it, Methuselah,' scolded Ellie cheerfully. 'You and I haven't exactly got one foot in the grave yet, I hope, so we've *all* got lives that'll be affected by coming here, and it can't do any of us anything but good – *especially* the boys and *including* us.'

She gave me a playful dig in the ribs, catching me off balance so that I toppled over sideways into an untidy heap on the inglenook floor.

'*Borracho! Ya borracho a la' nueve de la mañana?'*

I peeped round the corner of the bench seat to see old Rafael's grinning moon of a face at the kitchen door, flanked by the equally amused features of Sandy and Charlie.

'He's asking if you're drunk already – at nine in the morning,' laughed Sandy. 'That much I do know. But jeez, what language does this old guy speak? He's never stopped talking since we arrived at the orange trees and I've hardly understood anything. I just said "*sí*" every so often and he'd jabber on merrily for another few minutes. Talk about one-sided conversations!'

'Rafael's from Andalusia,' I grunted, hauling myself onto my feet.

'*Ah, Anda-loo-thee-ah. Mee pa-eeth. Qué pa-eeth tan fanta'tico!'* shouted Rafael in praise of his homeland, slapping the two boys simultaneously on the back and laughing like a drain until he coughed up another of his gob missiles which, in the presence of a lady, he elected to swallow lip-smackingly instead of launching it by spit-propulsion towards his favourite target over by the walnut tree. The age of chivalry would never be dead as long as old Rafael was still around.

Inspired by his show of good manners, I decided formally to introduce Rafael to Ellie, whom he had only ever seen once before – and that was when she was doing her own genteel charades version of Señora Ferrer's dogs dumping on the kitchen floor.

'Rafael, this is my *esposa*, Ellie. Ellie . . . Rafael.'

Old Rafael doffed his corduroy cap clumsily and offered Ellie an orange-stained, goat-tainted hand.

'*Encantado, señora,*' he whispered, bowing stiffly and allowing his little black olives of eyes to linger sparklingly for a moment or two on the obviously pleasing aspect of Ellie's ankles, knees and all points north.

'He's been picking up oranges full of worms and maggots, Dad,' confided Charlie with an expression of disgusted admiration.

'I can see that by the way he's looking at your mother, son,' I murmured knowingly.

'I – I believe you have lots of grandchildren, Señor Rafael,' Ellie stammered, freeing her hand from Rafael's sticky little digits and backing towards the makeshift Christmas tree. 'And, ehm – and I've got something which I'm sure you'd like.'

'Thank God he can't understand English, Ellie,' I said, shaking my head in disbelief. 'I've often accused you of overstating the bloody obvious, but –'

'I'm only going to give him some chocolates for his grandchildren,' replied Ellie indignantly. She turned round and rummaged under the tree. 'I've got an extra box or two here for just this sort of situation.'

While her back was turned, Rafael looked at me dolefully and tapped the side of his head with his forefinger. '*Un poquito mejor en la cabeza, la señora?*'

'Yes, she's all right in the head now, Rafael. *Completamente normal,*' I nodded.

Rafael wasn't convinced. It often happened that way, he stated, pursing his lips and inclining his head to get a better view of Ellie's upturned rear end. The good-looking ones are *loco*, while the best cooks, the best house-cleaners, the best goat-keepers, the best fieldworkers, the best breeders are all ugly bitches. He had gone for one of the latter variety himself. Not that he hadn't had his share of the other type, of course, but in the country, you had to be practical when it came to picking a woman to marry. *Hombre!* Why keep an expensive thoroughbred if she can't pull a plough as good as an old mule?

I was tempted to tell him that Ellie had actually cooked up a great *Potaje de Lentejas* the night before, but then thought better of it. No point in complicating the issue. I only hoped that Ellie hadn't understood any of Rafael's sexist remarks which, although normal and acceptable to him and most of his generation, would most certainly not have met with Ellie's approval. I knew that she didn't think much of the old, male-dominated Spanish domestic traditions, and I didn't want to see Rafael getting a tongue-lashing for a crime which didn't even exist in his book.

My mind was put at ease by Ellie's kindly smile as she handed Rafael the box of chocolates. She patted the back of his hand and Rafael melted – standing there looking dopily at his feet like Bashful of Snow White's Seven Dwarfs.

Ellie gently lifted old Rafael's chin with her fingertip and, looking tenderly into his eyes, she spoke softly to him in English: 'Merry Christmas, and I hope your grandchildren will enjoy these. Perhaps you might even deem it possible to allow your wife to taste one, you obsolescent little male chauvinist runt.'

Blissfully unable to understand anything of what had just been said to him, Rafael blushed and visibly turned to jelly. *'Mucha' grathia', señora.'* He hoped that *la señora* would find health, wealth and happiness in Mallorca and, above all else, would still bear for *el señor* more strong sons like these two. They were fine *muchachos*, keen workers. *La señora's* seed was good. Very rare in a woman of such . . . *actractivos.* She had hips of excellent child-bearing quality, he had noticed, and she should make full use of that gift now before she got too old. *Sí, claro qué sí.*

Ellie thanked him, smiling sweetly and this time absolutely clueless of what the blazes he had been on about.

I closed my eyes and mouthed a silent prayer of thanks to the god of language barriers. A peal of thunder rumbled around the valley. Message received and understood, I concluded. Amen.

'No charge for the oranges today, Rafael,' I said swiftly. 'Let's just call it an offering to the gods.'

Rafael looked totally bemused, but crossed himself all the same. *'Sí, grathia' a Dio', amigo. Grathia' a Dio'.'*

By the time I had given old Rafael's orange-laden scooter a push-start down the lane, the storm clouds had closed in and the rain had started to fall – just a few isolated drops at first, splashing down heavily on the ground like little liquid bombshells, peppering the dusty surface of the lane into a miniature moonscape of tiny craters. Another crack of thunder echoed down from the north, louder now and nearer, and in an instant, the heavens opened to release a deluge of water, whipped into face-stinging lashes by sudden swirling gusts of wind.

I made a dash for the house, but was already quite soaked before I reached the shelter of the kitchen doorway.

'Look at that rain,' gasped Charlie, while I stood shaking the water from my hair. 'I've never seen anything like it.

Look, it's bouncing about a foot off the ground out there. And I don't think your gift to the gods was appreciated, Dad,' he muttered out of the corner of his mouth so that his mother wouldn't hear. 'They're all up there peeing on you.'

'Very likely, Charlie. Very likely. But just get out of the way and let me in before I drown in it.'

Sandy was in the inglenook busily adding dry sticks to the fire. He looked round at my sodden state and started to whistle the first few bars of 'I'm Dreaming Of A White Christmas'.

'"Raindrops Keep Falling On My Head" might have been more appropriate,' said Ellie, handing me a towel. 'And I hope your old *compadre* gets a good drenching on his way home, too.'

'Why? And anyway, what did he say that bothered you enough to call him a male chauvinist runt of all things?' I enquired with a note of caution.

'It's not just what he said. To tell the truth, most of what he says is all gobbledegook to me anyway. But I don't like being weighed up like some heifer in a sale ring.'

Ellie stomped angrily across the kitchen to check on the progress of the Christmas lunch. She took a quick look inside the sizzling oven then slammed the door shut. 'Stuff him,' she snarled, and I guessed that she didn't mean the turkey.

A dazzling blue-white flash illuminated the gathering gloom of the day, followed almost instantaneously by an ear-shattering clap of thunder which seemed to explode directly above us, rattling the windows and shutters, and making the old house shake to its very foundations. The noise reverberated round the mountains, each cliff and rocky cleft acting as a sounding board, bouncing the deep rumblings from side to side across the valley.

Ellie came scurrying over and clung to my arm, burying her face in my shoulder. 'I'm terrified of thunder storms,' she whimpered. 'Please, *please* cover my ears and tell me when it's all over. I hate this. I can't even bear to look.'

'Closing your eyes and ears isn't going to do you much good if the lightning strikes the house,' said Sandy from the doorway.

'Yeah, you'll be dead meat like the rest of us,' grinned Charlie, 'so you may as well come over to the door and see the show. It's fantastic! Come on, Mum. Don't be such a wimp.'

Without warning, there was a sickening crunch, like a giant scythe hacking a swathe through timber, and I looked out to see a writhing snake of lightning slicing a fiery path down through one of the eucalyptus trees in front of the house. The lightning rammed violently into the tree's forked bough, tearing one massive limb away from the trunk as if it were the wishbone of a tiny bird, before finding earth and dying in a hissing fury of flames and vapour.

It seemed that our eyes had had but a split second to witness the unstoppable ferocity of the lightning before our ears were bombarded by the din of the accompanying thunder. It was as though we had been trapped inside the barrel of some huge cannon at the moment of firing. A deafening explosion engulfed everything. We felt the old flagstones shuddering beneath our feet. Earthenware dishes were shaken off their shelves and crashed to the floor, while the sound of broken glass from shattered window panes mingled with the ringing in our ears as the booming of the thunder echoed away down the valley.

For a few unreal moments, we seemed submerged in an eerie silence, but as our senses recovered from the shock, we became aware again of the battering torrential rain and the howling of the *Tramuntana*.

'Jesus Christ!' exclaimed Sandy, standing ashen-faced inside the kitchen door with his back pressed hard against the wall. 'That was a close one.'

I could feel Ellie trembling as she slowly released her hold on my arm.

'It's OK, dear,' I said reassuringly, though not quite able to control the nervous quiver in my voice. 'You can look now. It's all over . . . I think. We're all OK, that's the main thing.'

'But – but that terrible crashing noise and the breaking glass,' she stammered. 'I thought I heard breaking glass.'

'Aw yeah, Mum, you should've been watching,' piped up Charlie eagerly, scrambling out from his refuge under the table with sudden bravado. 'One of the big trees out there . . . completely knackered. And your plates. Thousands of bits. And all the windows in the house too.'

'How would you know, Charlie?' scoffed Sandy. 'You were under that table like a rabbit down a hole with a rocket up his bum.'

'Yes, don't exaggerate, Charlie,' said Ellie in a shaky voice. 'And don't you blaspheme, Sandy.'

'Bum isn't blasphemy,' objected Sandy.

'Right – perhaps the boys and I had better go and check the damage,' I suggested. 'Bring a brush and shovel, Sandy, and you see if you can find a couple of empty buckets, Charlie.'

I placed my hands on Ellie's shoulders and peered into her eyes. To my relief, they seemed already to have lost that look of total terror. Poor Ellie – she really was petrified of thunder and lightning, but at least the thunderbolt had cleared her mind of animosity towards old Rafael, and that was a blessing. Perhaps the old fellow's mechanical vote of thanks to the deity hadn't been quite so pointless as it seemed after all.

'You sure you're OK now?' I asked her quietly.

'Yes, I'm fine – honestly,' she assured me with a little smile. 'Off you go with the boys. I'll clear up this mess of broken dishes.'

Luckily, the sound of the glass falling onto the hard, tiled floors had been worse than the actual damage done. Only three windows had been broken, and the worst of the rain was easily kept out by simply closing the outside louvred shutters. All that could be done then was to stop up the draughts as best we could by pinning pieces of cardboard boxes over the missing panes.

'It doesn't look very tidy,' I admitted to the boys, 'but it'll have to do until we get a glazier out after the holidays.'

After the excitement and trauma of the lightning strike, a curious feeling of anti-climax seemed to prevail, followed by annoyance when we discovered that our telephone line had been brought down by the stricken eucalyptus branch, and near despair when we realised that our electricity supply had also been cut by the storm.

Without power, our electric water pump was out of commission, and without the pump, we had no running water at all in the house. I would have to get to a phone, but I knew that none of our near neighbours had one.

'Nothing else for it,' I decided. 'I'll have to set sail in the car.'

A perilous drive along flooded lanes strewn with mud slicks and boulders washed down from the surrounding high ground eventually brought me to the public telephone box at the far end of the village. After a repeatedly disconnected and totally frustrating conversation in my most painstaking Spanish with what turned out to be a faulty answering machine at the offices of G.E.S.A., the gas and electricity company, I was informed by the recorded voice that no domestic repairs would be carried out on Christmas Day. My problem would be listed and would be attended to in due course.

My call to the emergency number at the telephone company didn't even get a reply. *Feliz Navidad* indeed.

I returned home and broke the news to the others. The general mood of despondency deepened when I told them that all our water requirements would have to be drawn up by rope and bucket from the *aljibe*, the house's reservoir which was situated under the open terrace. But fetching water from outdoors during an incessant Mediterranean downpour would not be a pleasant chore, as we soon found out.

'Better go and stand by the fire,' I said to Sandy, who had drawn the short straw for the first trip to the *aljibe*. 'You've got to hand it to those old Mallorcans. They really knew what they were about when they built these huge inglenook fireplaces. A snug place to sit in when the weather's cold, and a handy place to dry off when it's wet. A grand idea.'

'An even better idea,' panted Sandy, standing shivering in the middle of the kitchen, 'though maybe not quite so grand, would have been a simple water storage tank up in the roof – like we have back home – so you don't have to go outside and get drenched every time the power goes off in a rainstorm.'

'You've got a point there,' I admitted sheepishly. 'I hadn't really thought about it before, but you're right. It's strange that they didn't put some emergency water storage in the lofts of these old houses.'

'What would have been the point when they didn't have electricity to pump the water up there?' Ellie reasoned.

'But that's no excuse now,' Charlie said. 'It's nearly as bonkers as having a TV aerial and no telly to watch.'

'Yes, I wonder why the Ferrers didn't put a tank in the loft when they had the electricity installed,' Ellie pondered.

'Too bloody tight-fisted, that's why,' I retorted. 'Anyhow, we'll have to make do with the rope and bucket method for now, so try to look on the bright side. Just tell yourselves that it's an authentic taste of the traditional way of Mallorcan country living.'

'Traditional? Antediluvian, more like,' Sandy mumbled, squelching off into the inglenook.

The ethereal bearers of seasonal goodwill still seemed to be giving our household a wide berth all right, and if the boys had been asked at that moment to name their most wished-for Christmas presents, one-way air tickets right off the island would surely have come top of their lists.

But Ellie was determined that our first Christmas at Ca's Mayoral was going to be a good and memorable one. She had made all the preparations, and no little inconvenience like a Mediterranean monsoon was going to dampen *her* festive spirit. And so she sat us round the big kitchen table where we pulled crackers, donned paper hats and went through the motions, at least, of being merry to the heartening sound of old Pep's logs hissing and sparking and giving off the sweet, sharp scent of rosemary which Ellie had sprinkled on the fire according to the Mallorcan Yuletide custom.

Her Christmas lunch menu was also borrowed from the rural traditions of our adopted country; a starter of *Sopa de Pilotes* – not, as Charlie presumed, the stuff that aeroplane drivers washed their feet with, but a brawny broth of little, herby meatballs, laid on a bed of oven-toasted bread. Then the turkey himself – a fine figure of a fellow who had spent his earthly days gobbling grubs and showing off his rear-mounted chief's head-dress to his modestly-feathered squaws whilst strolling through sunlit avenues of apricot and plum trees on a little *finca* further up the valley. It was, perhaps, a trifle ironic that the majestic bird had come to his final resting place on our Christmas table, stuffed with a porky mixture of chopped, dried fruit from those very orchards where he used to strut his stuff so grandly. But a shroud of Ellie's pomegranate sauce and a wreath of roast baby potatoes, baked tomatoes, peas and fried sweet potatoes paid colourful respect to this departed poultry potentate.

We rounded off the banquet with little bars of *Turrón* – a specially-baked-for-Christmas almond nougat from the local *pastelería* – and a palate-refreshing citrus custard made from old Maria's priceless eggs and the juice of our own lemons.

Ellie had done us proud.

The afternoon wore slowly on, and the torrential rain continued unabated. The thunder had receded steadily out over the sea towards Africa, and as a grey dusk descended early in the valley, we could see the distant glimmer of lightning flickering pink against the leaden clouds in the southern sky.

We sat on at the table, our paper hats gamely flaunting their gaudy colours against the deepening gloom, while outside the little terraced fields were like rice paddies, the soil carpeted by a layer of muddy water pock-marked by the falling rain. Somehow, the luscious green of the orange trees and their glowing, golden fruit seemed sadly at odds with such a dismal landscape. Even the tall palm trees standing like sentinels by the neighbouring farmhouses looked forlorn and alien, their exotic splendour wilting before our eyes in a wet and windy setting which belonged more to some bleak corner on the Atlantic seaboard of the Outer Hebrides than to the enchanting mountain hinterland of a Mediterranean island. The boys had been right; this hardly tallied with the popular image of a sun-drenched paradise valley.

'I'm fed up,' groaned Charlie, stretching himself lavishly and scratching his armpit. 'I think I'll go for a slash.'

'Whatever,' said Ellie, a look of blank indifference on her face, 'but just try to aim carefully this time. It's not nice.'

'Absolutely,' I proffered vaguely, calculating that the sun – if there had been a sun – would now have dipped below the yardarm of the western mountains and would have signalled, therefore, the advent of Happy Hour. 'Just aim carefully, Charlie. These tiled floors are lethal when they're wet.'

The gathering crepuscular murk was by now plunging the old kitchen in shadows of melancholy, casting a clinging web of nostalgia-tinged glumness over us all.

'OK – let's break out the candles,' I breezed. 'That'll brighten things up.'

'Why not?' replied Sandy, resting his chin on his fist and gazing vacantly towards the window and the inky, wet cloak of creeping darkness. 'To hell with the expense. It's Christmas.'

'Everybody's bored – that's the trouble,' Ellie said. 'There must be something we can do to cheer things up. Think!'

'I think I'll pour myself a large gin and tonic,' I said, wandering over to the cupboard. 'That'll certainly cheer me up. I've got a bottle of that Minorcan Xoriguer in here somewhere. Fancy anything yourselves?'

They shook their heads.

'Don't be such doom merchants,' I urged. 'Get the party spirit going here. A sherry, Ellie? A beer for you, Sandy?'

'No thanks, Dad. I'm still too full of turkey and things.'

'I'll pass as well, dear.'

'Suit yourselves,' I said, admiring the Xoriguer's old-fashioned stoneware bottle with its little round finger hole at the neck. 'You should try an orange juice with some of this in it, Ellie. You don't know what's good for you.'

Ellie shuddered and turned up her nose. 'I *do* know what's good for me, so no thanks.'

'Why won't the loo flush, Dad?' asked Charlie, stumbling back into the kitchen with a flashlight in his hand. 'I've pulled and pulled at the plug, but nothing happens. There's no water in the bog.'

'Exactly, young Mastermind of the year,' jeered Sandy. 'Watch my lips. There – is – no – water. Remember? No electricity, no water pump, no water?'

'Oh God, I completely forgot about that side of things,' I admitted. 'We'll have to cart in water from the *aljibe* for flushing the toilets too. Bring your torch, Charlie. Sandy and I will fill as many basins and pails as we can lay hands on. We may as well bring in enough water to last all night.'

Mercifully, the rain had begun to die out with the advent of darkness, and the wind had dropped to little more than a chilly, gusting breeze from the north. As we hauled the old wooden bucket up through the well on the terrace, we could see breaks opening in the clouds above the mountains, revealing little clusters of stars shining hopefully in the patches of indigo sky. The storm had passed, but our troubles had not, as we were about to discover.

'Take this pail of water upstairs and pour it down the loo, Charlie,' I said. 'Sandy and I will fill up the rest of these things. Oh, and don't forget to bring your pail back for a refill.'

Charlie was back in a trice. 'No way, Dad,' he gasped. 'I flung the water down the pan, but it hasn't worked. It just filled up and stayed there.'

'Yet another item disappears off the list of mod cons at Ca's Mayoral. Now we haven't even got a bog,' moaned Sandy.

'Not to worry,' I bluffed cheerily, 'let's get the plunger from under the sink. We'll soon get the loo working again.'

Fifteen minutes and a few hundred plunger strokes later, the pan was still full to the brim with water, and I was exhausted.

'Forget it,' I wheezed. 'There's no point in sploshing about any more in here. It's doing no damned good at all. I'll have to take a look outside.'

As is normal in rural areas, the house at Ca's Mayoral was not connected to a main sewer, simply because there *was* no sewer. Like most farmhouses, ours had its plumbing outlets piped into a nearby septic tank – an underground chamber in which the live-in bacteria digest and convert everything that enters their domain into a liquid which filters away through a soak-away system into the surrounding subsoil. A good septic tank, once working properly and not polluted with down-the-tubes supermarket bags, plastic ducks, toy cars, teddy bears and other non-biodegradable discards, should seldom, if ever, need attention. I had heard it said, though, that the increasing use of modern kitchen chemicals and washing machine detergents could upset the delicate digestions of the microscopic septic tank inhabitants, and might even ruin their appetites completely. But, then, I had also been told (by an old gamekeeper who was well up on such matters) that a dead cat, retrieved from the roadway and bunged into the tank, would soon have the armies of bugs feeling peckish and gobbling away heartily once more. It was all to do with advanced chemistry, the old boy had confided.

The gamekeeper's tip was not a practise that I had ever intended to employ, but as I trudged towards the septic tank on that wretched Christmas night, I was heartened, I confess, by the knowledge that, should the need arise, there was a good supply of potential bug-appetisers lurking about over at the Ferrer's *casita*. And the vicious, little, scurvy, feline devil that had lacerated my leg would do just nicely.

But Francisca's cats had no need to feel threatened on this occasion. When I shone my torch along the path towards the septic tank, the cause of the problem became all too apparent – and the solution did not lie in advanced chemistry. The tank was simply full to overflowing, the internal pressure having been sufficient to lift the concrete lid out of its bed, allowing a putrid porridge of sewage to bubble out over the ground.

'So what now?' Ellie asked when I broke the latest bad news to her back in the kitchen.

'As Sandy said, another mod con disappears off the list. Now we have no telephone, no electricity, no running water, and we can't use the kitchen sink, the toilet or the bath.'

'What a break!' Charlie enthused. 'No bath, and I can pee out of my bedroom window without getting a bollocking. Different class!'

'*And* you won't need to help with the washing-up,' Sandy added.

'Dead right, big brother. Yeah, this is turning out to be not such a bad Christmas after all.'

Ellie didn't greet the news so enthusiastically, however. 'I can hardly believe it,' she sighed, shaking her head slowly. 'And all that silly stuff we said about this old house having a good feeling about it. Well, I'm not so sure any more. I don't know . . . I really do wonder what it's going to throw at us next.'

I could see that the combined calamities of the day were beginning to get to her, and I dearly wished there was something I could say to cheer her up. But there wasn't.

'Now we can't even use the loo,' she muttered, her moist-eyed expression a picture of dejection in the dim candlelight.

That old feeling of gathering gloom was returning, deepened further, I feared, by the first little pangs of homesickness. As one, we were all beginning to feel a long way from home – stuck in a hostile old house on an alien island far, far away from the family and any friends who truly cared. Paradise lost . . . already.

Suddenly, our silent melancholy was interrupted by a timid knock at the kitchen door. On opening it, I was surprised to see old Rafael standing in the darkness with an oil lantern in one hand and what looked like a little tray of lumpy objects covered by a dish towel in the other.

He apologised meekly for disturbing us so late, *e'pethialmente* on Christmas night, but the word had got round the village that the storm had caused a power cut in the valley, and as he knew that our household possessions had not yet arrived from *E'cothia*, he had walked up to offer us the use of his lantern for the night. He only used it to check on his goats, he assured us, but they were all now safely inside and would need no further attention till the morning. Oh, and would we forgive him, *por favor,* for coming directly to the door, but the gate had been open and . . .

I put my arm round his shoulder and guided him into the warmth of the kitchen. 'Rafael,' I said, swallowing at the lump in my throat, 'you really shouldn't have. You've walked all this way on such a cold night just to –'

'U'ted tranquilo, amigo,' he smiled, placing his lamp on the table and patting the back of my hand. *'U'ted tranquilo.'*

The old man then turned towards Ellie, slipping off his corduroy cap and tucking it under his arm as he shuffled hesitantly across the room.

'*Señora,*' he whispered shyly, looking up into her bewildered face, '*mi mujer y yo . . . o sea, ehm . . .*' With unsteady hands, he held out the covered tray.

'Take it, Ellie,' I urged quietly. 'It's a gift for you – a present from Rafael and his wife.'

His wife had been surprised and delighted, *totalmente encantada*, by the box of chocolates so kindly given by *la señora*, Rafael stammered, nervously but carefully removing the towel. And would *la señora* honour them by accepting this humble offering in return?

Ellie's expression melted and her bottom lip began to quiver as she looked down at the group of tiny figures that old Rafael had uncovered. She was clearly overcome.

'*E' un belén, señora,*' he stated, a look of apprehension furrowing his gnarled brow. '*E' el belén ma' viejo de mi familia.*'

It was a miniature depiction of the Nativity, a *belén* or Bethlehem, a model scene which is given pride of place as the most cherished symbol of Christmas in every Spanish house at this time of the year – and Rafael's gift was the oldest *belén* of his own family. The base was the lid of an old round biscuit tin on which the interior of a simple stable had been crudely fashioned out of clay, cracked now with age, the once-bright yellow, green and red paint faded and flaking. The centrepiece was a little manger, roughly carved from a scrap of firewood, the head of the infant Jesus little more than a dab of paint amid the shreds of straw that were his bed. In adoring attendance stood Mary and Joseph and the three Kings, their bodies merely little columns of clay with almond twig arms, their bowed acorn heads adorned with fuse-wire haloes. Huddled at a reverent distance were three tiny animals of nondescript shape (presumably sheep), and nearby, a larger creature vaguely resembling a unicorn, but more likely to be a donkey that had long since lost one fragile clay ear. Completing the scene was a backdrop of jagged mountains with pine cone trees dotting their slopes, and a match stick star covered in silver tobacco paper shining down from its heavenly orbit atop a bent metal knitting needle.

'*La señora* does not like my *belén*?' Rafael pleaded, turning to me for interpretation as the tears started to trickle down Ellie's cheeks. *Hombre*, he had made this *belén* with his own hands, he stressed in high dudgeon. *Sí* – all those years ago! *Bueno*, perhaps it was not worth countless *milliones* like the grand jewel-studded, silk-clad *belén* in the Palacio March in Palma, and maybe his *Jesús, Maria y José* did not look as neat and tidy as those plastic, store-bought figures that people put on their fancy *belénes* these days, but his little *belén* was unique; it was part of him. *Madre de Dios*, whose spit did we think had been used to stick all those pine cones to the mountains? '*Cuarenta putas!*'

Quite what his last exclamation – meaning 'Forty whores!' – had to do with the Nativity, I didn't know; nor was I about to ask. Rafael was in full, furious flow now, so I took a calming sip of Xoriguer gin while he continued his tirade – my offer to him of a glass of the same having been brushed aside with a sweep of a callused hand.

'*Córcholi'!*' he snapped, his voice rising to a gnat-like whine, his cap planted firmly back on his head. 'That *belén* is not just e'*pethial*, it is sacred, holy, a product of my very soul!'

I tried my best to translate all this to Ellie and the boys, while Rafael took a well-timed break and stood grimacing and shaking in an effort to stem his rising temper.

'*Ah sí*, my first-born son,' he continued at last, his composure apparently restored and his tone more reflective. 'I made that *belén* for my first-born son . . . for little Rafaelito.' He lowered his eyes, sniffed loudly and drew the back of his hand under his nose with a dramatic slurping sound.

I looked over at Ellie, and I guessed from her reddening eyes and trembling chin that she was getting the gist of this highly emotional performance.

'*Rafaelito, oh mi Rafaelito,*' the old man whimpered, pausing to cross himself with a quivering hand. '*Sí, señora,*' he continued, raising his dewy eyes to Ellie who was now biting her lip and trying hard to hold back the tears, '*sí*, my little Rafaelito was born on this day also, and I made this *belén* for him . . . all those years ago.' The old man took a deep, shuddering breath, his chin dropped onto his chest, and a big tear fell from beneath the peak of his cap and landed with a little splash on the flagstone floor. 'Today is the birthday of my *Rafaelito. Ah sí, amigos* – today he . . . he would have been . . .'

But Rafael was unable to go on. He lowered his head, his stooped shoulders began to heave, and he raised his hands to his eyes as great, silent sobs racked his frail old frame.

Ellie hesitated for a moment, pressing her fingers to her lips; then, unable to contain herself any longer, rushed to old Rafael in a flood of tears, throwing her arms around him in a smothering, motherly hug. 'There, there, there,' she soothed, patting the top of his corduroy cap and rubbing his back like a baby. 'Your *belén* is the nicest one I've ever seen, and we'll treasure it always – I promise.'

'*Grathia', señora. Oh, mucha' grathia',*' crooned Rafael, not understanding anything that Ellie had said, but cuddling in all the same. *Sí*, it had been our two sons who had reminded him of Rafaelito when they were picking oranges with him yesterday, he explained. 'Such fine *muchachos* – just like Rafaelito when . . . but the *difteria* . . .' He cleared his nose in a loud inward snort, swallowed hard, then went on: 'And that is why I want you to have the *belén* that I made for Rafa . . . Rafa . . . *oh, mi Rafaelito-o-o!*'

A spasm of shivering ran through Rafael's body, his knees began to quake, and he started to bawl like an infant – though his cries were muffled considerably by Ellie's bosom, which just happened to be at face height for our grieving goatherd. With one hand he clung to this bountiful source of refuge and comfort like a koala clamped to a gum tree, while the other hand eagerly reciprocated Ellie's back-rubbing motion.

This touching act of mutual solace and gratitude continued to the sound of muted blubbering until Ellie decided (perhaps not entirely without justification) that Rafael's back-rubbing hand had wandered perilously close to the bum-groping zone. It was then that her natural sense of compassion turned instantly to one of vilification, and she swiftly disentangled herself from Rafael's roving paws with a shocked giggle and an indignant, warbled scream of: 'YOU DIRTY OLD PERV!'

Rafael stood scratching his head through the cloth of his cap, his dumbfounded expression confirming that, in his mind, *la señora* was, indeed, a couple of oranges short of the full kilo.

After much Andalusian huffing, puffing, arm-waving and lisped protestation, Rafael was eventually placated and persuaded to stay a while and join us for a traditional Mallorcan Christmas supper of *ensaimadas* and mugs of thick, hot, sweet chocolate – albeit that he and Ellie were seated in mutual mistrust on opposite sides of the inglenook. His precious *belén* was installed with due respect and ceremony on the most prominent shelf in the kitchen, where the primitively sculptured characters were illuminated by clusters of birthday cake candles ingeniously stuck into two half potatoes by Sandy and Charlie. Blessed are the peace-makers.

That unforgettable day drew to a mercifully peaceful close, with a Xoriguer-mellowed Rafael snoring blissfully by our familial fire, the heat-enhanced hum of goats rising from his trousers and lending an authentic ambience to the little clay stable glowing proudly on the shelf above his head. Even Ellie had to concede a contended smile.

As Charlie said, it had turned out to be not such a bad Christmas after all.

– SEVEN –

THE FLYING CHIMNEY SWEEP

We looked on as if spellbound while the engine of the little tanker truck from Osifar, the sludge disposal company, chugged away soothingly in the warm air, its pulsating flexi-pipe greedily gobbling up the vile stew from the cavernous darkness of the septic tank like the trunk of some depraved mechanical elephant at its favourite watering hole. The Spanish term for septic tank is *pozo negro*, or black well, and under the circumstances, I could not have thought of a more accurate description.

There were only the three of us – the Osifar operative in his overalls of sensible brown, myself, and a spectacular pink hoopoe bird which had flitted through the orchard on its black-and-white striped butterfly wings to alight inquisitively at a favourable vantage point in what was left of the lightning-struck eucalyptus tree.

'When do you think this *pozo negro* was last emptied?' asked the Osifar man hypnotically.

'Señor Ferrer who sold me this house says that the tank has never been emptied in the ten years since it was installed. I spoke to him about it only this morning. Never needed to be emptied, he said.'

The Osifar man just shook his head and pulled a one-shoulder shrug. He obviously couldn't have cared less. You've emptied one septic tank, you've emptied them all, was his attitude, I supposed, and if he hadn't needed his meagre wages badly enough, he wouldn't have been standing there on Boxing Day making small talk and watching other people's accumulated excrement being sucked out of a smelly hole in the ground.

'Yes, Señor Ferrer even showed me the original plans for this *pozo negro* this morning,' I said, trying to generate some spark of professional interest in the subject. 'There's another chamber after this one, full of rocks to act as a kind of filter. Then what's left – just dirty water really – is piped away into a soak-away pit in the field over there somewhere.'

The Osifar man shrugged his other shoulder. The hoopoe bird raised its fan-like crest into an instant Mohican head-dress, lifted one leg in the air and proceeded to preen his crotch acrobatically with his long, curved beak. He couldn't have cared less either. The sludge pump chugged on regardless.

Apart from the transient stench in the immediate vicinity of the septic tank, the day had turned out fine. The sun was shining again, the sky was blue, and although a few wispy clouds were still being wafted southward over the mountain tops by the dying gasps of the *Tramuntana*, everything in the valley was lovely. Not a breath of wind disturbed the orange leaves, glistening dark green and shiny clean after the rains, and not even a ripple of air moved the patchy banks of mist lying ankle-deep among the orderly platoons of fruit trees. It was shaping up to be one of those beautifully pleasant Mallorcan winter days – apart from the transient stench in the immediate vicinity of the septic tank.

'Maybe the pipe connecting the *pozo negro* to the filtering chamber is blocked,' I suggested in a further attempt at technical conversation. 'Perhaps that's why the *pozo negro* was overflowing.'

The Osifar man didn't even bother shrugging this time. He only raised a sceptical eyebrow and shuffled, hands in pockets, over to the septic tank. Leaning forward, he peered into the dark void for a minute, ostensibly oblivious to the choking reek rising into his nostrils, then looked back at me and did a combination head-shake, shrug and double eyebrow-lift. 'No outlet pipe here,' he said, 'so no filter chamber and no soak-away pit either. *Coño*, what you have here is a simple storage tank. *Nada mas.*'

'But the plans,' I protested. 'Señor Ferrer explained them to me and they show a complete three-chamber system – *totalmente comprensivo*. It's all there in the plans.'

'*Los planos?* Forget them. This *pozo negro* was built the cheapest way possible, and there is no more to it. *Los planos* mean nothing. *Nada.*'

'But how could the public health people allow the plans to be completely ignored?' I asked in dismay. 'How could they approve a useless pit like this?'

'Approve? Approve?' The Osifar man stepped back a pace and looked at me aghast. '*Hombre!*' he murmured, shaking his head in a mixture of disbelief and pity at my naïvety.

My heart sank as the glaring truth struck me. Tomàs Ferrer had obviously used his elevated position with the local authority to have this cheap botch-up conveniently ignored by the relevant official inspector.

'So, what now?' I asked, stunned and sickened.

'So now, with normal family use, you will have to book us to empty this tank for you about once a month,' advised the Osifar man impassively, handing me a bill for the cost of today's call-out.

'But how did the Ferrers manage without emptying it for ten whole years?' I queried, quickly working out that I faced an annual outlay of over a thousand pounds to save our little piece of paradise from being slowly submerged in our own sewage. 'I mean, they couldn't have been constipated for that long, could they?'

'Not on a fruit farm anyway,' responded the Osifar man with an unexpectedly wry turn of humour. We were evidently touching on his favourite subject at last. 'Hombre! They would only use the house toilet if they were really caught short. The rest of the time, they would prefer to do what they always did before they had a pozo negro. I go round these country places all the time and I know what paisanos are like. If it is going to save a peseta or two, they would rather go into a field and do what comes naturally behind a wall. Remember, a handful of fig leaves is cheaper than a toilet roll, no?'

He let out a dirty chortle and started to reel in the flexi-hose. Another mission had been accomplished, and what was more, it had finished on a happy note. His day was off to a flying start.

I couldn't share his mood of elation. 'Now I know what it feels like to be a handful of fig leaves,' I thought glumly. The Ferrers had done it to me yet again.

'*Poop – poop!*' hooted the hoopoe bird, flashing his Mohican cut in distinctly derisory style before gliding off jerkily in the direction of the Ferrer's *casita*. '*POOP – POOP – POOP!*' he repeated.

'Very apt,' I muttered. 'Even the bloody bird's rubbing my nose in it now.'

'Was that a cuckoo I heard just then?' called Ellie, hurrying excitedly out of the kitchen.

'Not exactly, but the next best thing,' I grouched. 'In any case, it was probably some kind of winged messenger working for Tomàs Ferrer. Nothing would surprise me now.'

'Winged messenger? You mean he's keeping pigeons over there? You've lost me. What message are you talking about anyway?'

'The message, Ellie, is that the septic tank is not quite the efficient sanitary filtration and disposal installation that Ferrer showed me on the plans this morning. The only disposal work this so-called septic tank will do well will be to act as a permanent drain on our financial resources. That's the message.'

'Mm-mm, I get it. First the cats and dogs, then the cooker, the washing machine, the water boiler, the entire electrical system . . . now this is going to cost us more money, isn't it? We've been put upon yet again by the Ferrers, haven't we?'

'Yes, but "put" isn't the word I'd use in this particular context,' I said bitterly.

The Osifar man had been silent for a while, but he hadn't been idle. He had been thinking, he told me, and perhaps he had *una solución* to *el problema*. Too much liquid – that was the thing. If we could only keep the *pozo negro* free of the excess liquids which would build up quickly in it from the bath, washing machine, sinks and so on, it might still function properly. If we could just control the water level in the tank, the bacteria living in it would then be able to break down the solids as intended in all of these systems, he reasoned. After the bugs had done their work, the resulting sludge would mix with any remaining liquids for disposal. *Comprende?*

'But how can we dispose of any of these liquids without still having to hire you?' I asked. 'There's no outflow pipe, you say, and without a filtering chamber, the liquids will still be foul, surely.'

'*Correcto.* And that foul water is the precise key to my *solución, señor.*'

I looked at him blankly.

Because the liquids were unfiltered, he explained, they would still contain an abundance of nutrients. The high value of human effluent as an agricultural fertiliser was well known in his business, and if we followed his advice, we would have a self-replenishing reservoir of the precious brew which we could pipe out to wherever it was most needed on the farm simply by fitting an electric pump to the *pozo negro*. *Madre mía*, during the long summer droughts, such a free supply of liquid fertiliser for irrigation would be a godsend. The *pozo negro* would be saving us money instead of costing. '*Una solución perfecta, no?*'

'Eh . . . yes. I follow your logic, and it's a great idea – no doubt about it. But what about the smell? That stuff will stink to high heaven, won't it?'

The Osifar man winked cunningly. '*No problema* – provided you pump it well out of smelling distance of your own house. You see, I also have a small *finca, señor*, and I know from experience that the *pozo negro* waters can be excellent for growing tomatoes.' The corners of his mouth inched into a mischievous little grin. 'And I also know from experience that, looking around your *finca*, the best position for growing tomatoes here might well be over there – right beside the weekend *casita* of your neighbour, Tomàs Ferrer. *Me entiende?*'

I could feel the clouds of despondency dispersing already. Oh, the simple, sweet, vengeful beauty of it all. What a wicked little gem of an idea. Not only would our sewage problem be solved, but the Ferrers would get their own back – or worse – all summer long, and we would get a bumper crop of free tomatoes. Better still, the Ferrers couldn't even complain about the smell without risking our exposing the corrupt details of their sub-standard septic tank scam.

'*Amigo*, you are a genius,' I beamed, counting pesetas into his hand. 'And here's extra for yourself. You're an absolute genius. *Feliz Navidad!*'

The Osifar man accepted my plaudits and the tip with a gracious nod of the head. 'Just do not tell anyone that I gave you this advice, *señor*. If my boss found out that I told a customer how to avoid paying Osifar to empty a *pozo negro*, it would land me right in it.' He took a sidelong glance at the open septic tank. 'And in my occupation, that is a punishment best avoided.'

Maybe the old house's spirits of times past finally decided that we'd had enough misfortune hurled at us for the time being. I didn't know – but after the traumas of that stormy and lavatorial Christmas, events at Ca's Mayoral soon took a decided turn for the better.

On the other hand, perhaps the news of our unseasonable tribulations had simply triggered a feeling of sympathy in the local community, where the country trait of helping a neighbour in distress still prevailed – by and large. Whatever the reason, all our most pressing problems were attended to without undue delay.

The telephone line and power supply were quickly restored, and Juan the plumber/electrician gave us his undivided attention until he had completed his upgrading of our electrical system and had rigged up an electric pump at the *pozo negro* for future irrigation and neighbour-annoying purposes.

Even Juan Juan the carpenter man turned out right away to fix our broken windows, and he removed the shutters which needed repairing with a promise that they would be returned in good-as-new condition within *dos o tres días – máximo, máximo*. And sure enough, he was back with our refurbished *persianas* in his little Renault 4 van exactly three days later.

It all seemed too good to be true in this land of *mañana*, where two or three days could mean anything from next week to next year or even never.

Juan Juan, however, was about to reveal his particular motive for punctuality. Sliding the final shutter from his van, he brushed off some flakes of sawdust with the back of his hand and predicted that these *persianas* would now last for decades – provided, *naturalmente*, we gave them a good coat of paint before re-hanging them. *Sí, muy importante, la pintura*. Nervously, he cleared his throat and shifted his weight from one foot to the other, looking at me and his fidgety feet in quick succession.

At first, I thought that perhaps he needed to use the toilet and was too embarrassed to ask, but then it dawned on me that it was more likely that he only wanted immediate payment for a good job promptly done.

'*Lo siento*, Juan,' I said apologetically. 'Of course, I must pay you for the windows and shutters right away. Just tell me how much I owe you, please.'

No, no, no – there was no hurry for that, he assured me. His wife would prepare the bill in good time. Maybe even next month. She was always in a hurry, his wife. Mallorcan, of course. He himself was from Ibiza. Much more relaxed, the *Ibicencos*. *Sí, sí – mucho mas plácido*. It was the Arab in them, *naturalmente*.

I nodded keenly in tactful agreement and waited for some clue to the reason for his apparent uneasiness.

The little *carpintero* scratched his head of curly, grey-flecked hair, then plucked nervously at some wood chips which were enmeshed in the sleeve of his woolly cardigan.

He hoped he was not being *impertinente*, he implored at last, but our neighbour, old Jaume, had mentioned to him some time ago that we might be interested in buying a tractor, and as he himself owned just such a tractor, he had thought that . . . But no, no – if I had been truly interested, I would have said so before now, he was sure. He had been too presumptuous, and would I please forgive him? '*Perdón, señor. Eh, usted perdone.*'

Juan Juan turned away quickly and closed the doors of his van, clearly regretting that he had finally plucked up the courage to broach the subject of the tractor. He came forward again, rolling his head awkwardly and looking dejectedly at the ground. I sensed the delivery of more apologies.

'Look, Juan,' I laughed, 'it's I who should be apologising to you. I really am sorry, but what with the storm damage and everything, I totally forgot about your tractor.' I gave him a reassuring slap on the shoulder. 'Yes, I'm interested in it. Yes indeed. *Ciertamente!*'

The expression on Juan's face was instantly transformed into one of undisguised relief. The *señor* was very kind, he said smiling shyly. *Muy amable*. He had really expected me to ask about his tractor when I contacted him about the house repairs, but when I said nothing, he assumed that I wanted to wait and see if he was a man of his word before doing any more business with him. *Es normal*.

I was certainly glad that Juan Juan's reasoning had prompted him to be expeditious, but more importantly, it was unusually gratifying to find someone who still cared more about ethics than trying to push a fast deal. I liked this little chap. Where would it be possible to see his tractor, I asked?

Juan explained that, like many tradesmen who lived and worked in towns like Andratx, he also had a *finca* quite a distance out of town. His farm had been passed on by his wife's parents, and although it was very small, he enjoyed working on it as a *pasatiempo* at weekends. It was good for his children too – a healthy place high in the mountains where they could play safely and sample the simple life of their ancestors for a day or two each week. *Muy importante, esto.*

I agreed that it certainly was important for children to learn about and value their rural heritage. It was fortunate that so many town-dwelling children in Mallorca could still experience that, thanks to the surviving tradition of inheritance of the small family farm. If only such a system had survived in Britain, I added.

So, this tradition did not exist in the *señor's* country also, asked Juan Juan, looking genuinely surprised.

'No, I'm afraid not,' I replied. 'Farming has become more and more of an industrialised business in *Gran Bretaña*. Families can't make a living on the smaller farms any more, so they're being bought up by the big farmers. And that's the way things have gone over there; fewer farmers and fewer farms – but bigger farms, more efficient farms, more mechanised, requiring less workers. It's in the name of progress, you know.'

The little carpenter seemed truly saddened. How could it be progress when the countryside was being stripped of its life, he wondered? Without successive generations of families keeping their feet on their own land, the countryside would be a dead place – a desert. A thousand efficient machines could never replace the joy and satisfaction that a man drew from hearing the laughter of his children playing in the fields as he tended his *finca*, no matter how small it might be.

'So, I gather that this tractor of yours is up at your *finca*. But when can I see it?' I asked as we walked across to his van.

He glanced at his watch. 'Why not right now, *ahora mismo*? It will only take half an hour, if you can spare the time.'

I slipped into the van beside him and we set off on the twisting mountain road towards Capdella, leaving the terracotta tiled roofs of the farmhouses and the little fields of fruit trees looking like matchboxes on scraps of striped cloth far below.

As we reached the highest point on the main road, Juan Juan steered his van sharply to the right and into a narrow cart track which zigzagged and bumped even further up the mountain, climbing higher and higher into a rugged world of pine forests and mighty cliffs, of boulders and rocky screes washed down the sheer slopes by flash torrents and waterfalls born of countless *Tramuntanas*. My ears began to pop and the little van rattled and bounced onwards and ever upwards.

Through the open window, I could smell the damp sweetness of the mountain air, heavy with the fragrance of pine resin and heather. High above the peaks ahead of us, a red kite floated on motionless wings, spiralling lazily heavenward on the invisible currents of warm air rising up from the valley floor.

'The lord of the skies,' said Juan Juan reverently, pointing towards the soaring bird. 'At this very moment, he is looking down and pitying us poor mortals struggling here on the ground, while he drifts so effortlessly above us on his wings of an angel. *Es magnífico.*'

And I had been thinking to myself that the kite was only up there to keep his beady, telescopic eyes peeled for a nice, plump, little Earthbound rodent which he could snatch up and rip to shreds for lunch. But on balance, I had to concede that Juan Juan's romantic Latin interpretation was slightly more in sympathy with the classic sylvan scene that lay before us.

I bowed to his gentle poetic observation. 'You're right, Juan. He is a magnificent bird.'

'Sí,' said the carpenter, straining to look over his shoulder into the back of the van, 'and if I had my gun with me, I would shoot the brute here and now. He is the one that swooped down and flew off with my little girl's pet rabbit last weekend.'

'Even lords of the sky have to eat,' I suggested philosophically.

'Correcto. But of all the rabbits in the forest, why did it have to eat that particular one?'

'I'm very sorry, Juan. I didn't intend to appear insensitive. I can understand that your daughter must have been very upset indeed.'

Juan Juan's face was expressionless. 'We were all very upset, señor. The damned rabbit was almost ready for the pot. Bastardo bird!'

He stopped the van and left me to ponder the complexities of his attitude towards nature, while he went to open a small wooden gate that barred our way.

We had reached the end of the mountain track – a clearing in the trees where the steep gradient levelled out into a curved wedge of carefully tilled land. As we drove into this remote little farm, I could see more narrow banks of terraced fields clinging to the mountainside, one on top of the other like the steps of some ancient ruined temple, the precious slivers of deep soil retained against the erosive forces of rain and gravity by meticulously-maintained drystone walls.

Tucked cosily into the hillside stood the farmhouse – in reality little more than an old stone shack with a mono-pitched roof clad in faded clay tiles, and with a lean-to shed propped against either gable. Yet it was plain to see that Juan Juan and his family took great pride in this little example of Mallorca's rustic past. The front door and the shutters covering the window were freshly painted green, and the random, honey-hued stonework had been painstakingly picked and pointed – all the way up to the chimney which protruded through the roof, crowned with a rainproof 'hat' of matching tiles.

In stark contrast to the surrounding mass of mountain and forest, everything about this lonely *finca* seemed in miniature, the terraced *bancales* with their scatterings of almond and olive trees dwarfed by the imposing, untamed bulk of landscape which appeared to me to tower threateningly on all sides. At first, it struck me as strange that people could ever have chosen to live, work and bring up families in such an isolated place, where even a trip down to the village in the days before motorised transport would have meant an uncomfortable all-day journey by donkey and cart.

'A very beautiful place, my little *finca*, no?' said el *carpintero* while he fumbled with the padlock on the door of one of the sheds.

'It certainly has a charm all of its own,' I replied.

Juan Juan recognised that I was only being polite, and he smiled knowingly. 'I understand, *señor*. When I first came to this place, I also felt that it was too far off the beaten track, and I did not like the mountains looming over me. I had never known such wild places in Ibiza.' He made a sweeping gesture with his hand and looked round at the surrounding scenery. 'But now . . . I have come to appreciate that this is perhaps the nearest place to heaven that I will ever know. I love it. *Es celestial.*'

Realising that I didn't exactly share his feelings of adoration, Juan took me by the elbow and guided me towards the far end of the little field where it curved sharply round the side of the mountain. We turned the corner, and my breath was quite literally taken away by the view which suddenly unfolded.

The land fell away sharply in front of us, and I hesitantly approached the low retaining wall of the *bancal*, peering charily over its parapet into the dizzy depths of the valley. It was all just as old Maria Bauzá had described. Other small *fincas*, most of them unable to be seen from below, clung to the mountain or perched precariously on ledges and ridges, surrounded by tiny strips of terraced land, many of the ancient walls now in ruins.

Through the traces of mist which still hung over the valley bottom, I could make out the miniaturised form of the house and buildings on our own farm, from this great height seeming so much closer to the *fincas* of our neighbours than they actually were.

Far away to the left, the ochre rooftops of old Andratx could just be seen huddled tightly and securely round the church, their very presence all but concealed by the evergreen foothills rising and falling in craggy undulations at the base of the mountains as they rolled down towards the coast. Although the far distant air was hazy with moisture being drawn up by the heat of the midday sun, I could still spot a few of the villas above the Port of Andratx – tiny specks of white dotted about the pine-clad slopes, with the rich blue of the Mediterranean Sea fading and merging into the horizon beyond. This was the red kite's eyeview of our valley, and it truly was a stunning sight.

'Did you ever look down on such a *vista*, *señor*?' asked Juan proudly.

I had to confess that I had not, and I could say no more. I could only stand there and gaze in silent admiration.

'Ah, but if you are impressed by this now, you should see it in summertime. I often come and stand here on the long, late evenings and watch the sun setting behind those mountains over on the other side of the valley. We still have sunshine up here a long time after the town and farms down there are cast in shadow. And on the winter mornings too, the sun rises between those two peaks in the east and it seems to shine on our little house before anywhere else. *Sí, sí* – whoever picked this spot for his *finca* all those centuries ago was a wise man indeed. *Claro qué sí.*'

We turned and started back towards the house.

'And the air up here,' added Juan Juan breathing deeply, '. . . ah, it is always so fresh and cool, even when the terrible heat of July and August is torturing everybody down in the valley. I tell you this, *señor* – if I did not have to attend to my business in Andratx, I would live up here all summer long.'

The little carpenter stopped and touched my arm. 'Be silent for a moment and just listen. Then tell me, what do you hear?' He stood quietly beside me, looking up at the mountains and smiling contentedly to himself.

'Ehm, I hear nothing at all,' I replied, slightly bemused, 'except a few birds chirping . . . and the breeze in the pine trees. But nothing else really.'

'*Exacto*. Apart from the things of nature, this *finca* is an empty place now – a wilderness. But at weekends, it all comes back to life again. Once more there is the happy sound of children at play, there is smoke rising from the chimney, the smell of food cooking in the kitchen, and I can savour it all as I work in the fields. *Es celestial*. I feel sad that the little farmers in your country have lost all this, *señor*.'

I nodded my resigned agreement and we continued on our leisurely way back to the farmstead.

Looking back at it from this direction, the little house now seemed transformed into a picture of quiet idyllic charm, the mountainous backdrop exuding an air of protective serenity, the slopes reflecting the warm light of the southern sun while providing perfect shelter for the tiny farm when the winter storms blasted their fury down from the north. Juan was right; the original *campesino* had chosen this spot wisely.

'I'm beginning to see what you mean, Juan. This really is a heavenly place. But the remoteness – I mean, when people lived up here permanently, how did the children get to school, for example?'

'School?' laughed Juan. '*Hombre*, this was their school – the mountains, the forest, the fields. Maybe their parents could teach them to read and write, maybe not. In those days, such things were not important. To survive was the only important thing, and they do not teach you in school how to stay alive on a place like this.'

'But how *did* they survive? It's such a small amount of land to live off. How could a man and his wife manage to raise a family here?'

'I often wonder about that myself, *señor*, and I can only imagine that it could not have been easy – not if we judge it by today's standards. And yet, they would have had almost everything which they needed . . . except money, *naturalmente*. They would have kept hens, a few rabbits, some sheep and goats for meat and milk – none of these animals require much water, you know – and perhaps even a pig or two to root about in the forest, which would also provide rough grazing for the sheep and goats.' Juan Juan thought for a moment. 'In addition, they would grow oats and beans and some vegetables between the trees on the *bancales*. There would have been many more trees then – all almond and olive, you understand, because they need no water except what they can draw from the earth through their roots. You never need to irrigate almonds and olives, and that is why they thrive in such conditions.'

'So what about water up here? Where did they get it from?'

Again, Juan Juan echoed the words of Señora Bauzá. 'From the sky. The rain fell onto the roof and down into the *aljibe* under the house, just as it does today.'

'But surely the winter rainfall wouldn't be sufficient to keep a family and their animals supplied all year long.'

'Perhaps not – depending on how dry or wet the winter had been. But as ever, they could always rely on the mountain to sustain them. There are a few places where the water springs out from the rocks at all times of the year – clear, cold and sweet. The *campesinos* knew those places well and would carry barrels or even goat skins full of water from these *fuentes* when they needed it. In some parts, they even built *canaletas*, little canals lined with stone, to channel the water to the lower *fincas* for irrigation, or even to drive water mills in the valley. *Sí*, for them, the mountain was a friend – a provider of firewood to keep them warm in winter, and of many things to eat, like pigeons, partridges, rabbits, wild goats, mushrooms, berries, herbs and wild asparagus and, above all, water.'

'But what about things like tools and clothes? With little or no money, how did they come by things like that?'

'*Mi amigo*, all such *problemas* could be solved by a trip to the weekly market in the town. Today, the Andratx street market may seem little more than an attraction for thousands of *turistas* every Wednesday, but not so long ago, it was the nerve centre of rural life in areas like this. On market days, the farmers from these mountain *fincas* would set off at first light with their donkeys and carts loaded with anything which might be surplus to their family's requirements – a few eggs, a chicken or an old hen, a lamb, a piglet, the food which I have told you they could find in the forest, even a bucket or two of snails if there had been a shower of rain. Then, in the autumn, they would have their freshly harvested almonds and olives to sell, and they would trap as many migrating thrushes as they could, *naturalmente* – two or three million every year, they say. A great delicacy in Mallorca, the thrush.'

'*Naturalmente.*'

'Of course, during spring and summer, a man might become a *carboner*, a charcoal maker, and move into the forests with his whole family, living in little stone huts with roofs of branches and reeds, and building *sitjas* – mounds of chopped up holm oak covered with green wood and earth, then slowly and carefully burnt to make the charcoal. In the days before *butano* gas, charcoal was the main fuel for cooking and heating on the island. *Sí*, there was much demand for the product of the *carboners*. But not much money, *naturalmente*.'

'*Naturalmente*.'

Juan Juan pointed north towards the towering summit of Mount Galatzo. 'And in the highest mountains of the Sierra, a few *campesinos* might work as snowmen in winter.'

'Snowmen in Mallorca?'

'*Sí* – *nevaters*, they were called – snowmen. They had stone-lined pits, or snow houses, dug into the ground up there, and when the snows came, they would shovel it into the pits, tramp it down, continue in this way until the pits were full, then cover them with ashes and reeds until summer.'

'What then?'

'Then, the *nevaters* would extract the compacted snow and cart it to the villages and the city to sell for making ice cream or for medicinal purposes. That was before the days of freezers and refrigerators, *naturalmente*.'

'*Naturalmente*.'

'So you see, those people knew how to survive on these high *fincas*, and they knew how to make good use of all the gifts that the mountains and forests had to offer. They had many skills, the mountain *campesinos*, and they had to work hard to make a living, I am certain. We are all happy to carry on the traditions of these small farms, of course, but I am glad we only need to do it as a hobby, a *pasatiempo*.'

'And those old-timers had to be businessmen too.'

'*Correcto*. They would sell or trade their own wares for the things that they needed – a way of business as old as time, and it worked. And if a *campesino* had a good day at the market, perhaps there would be a few pesetas extra in his pocket for a drink or two to accompany a good yarn with some friends in one of the bars. And if he had had a *very* good day of trading, his *amigos* could always be relied upon to bundle him into his cart in good time for his mule to deliver him safely home to his wife in the mountains before nightfall. *No tan malo, no?*'

'Not so bad at all, Juan. In fact, it reminds me of a few old farmers I used to meet at the cattle markets in my own country, except they had no sensible mules to look after them. They drove themselves home in their cars, no matter how much they'd over-celebrated a good sale, or had over-mourned a bad one. They always seemed to make it, though.'

'It must have been the way on market days everywhere, *señor* – before progress made it necessary to invent the *borracho* bag.'

'Yes, and on the subject of the "drunk" bag, the breathalyser, I see that you have some grapevines, Juan,' I noted as we approached the house. 'For making your own wine?'

'No, no. I have no time to make wine, *señor*. No, I prefer to buy it from the *supermercado*.' He gestured towards the little angle-iron pergola in front of the house. 'These vines are only to provide some shade when we eat outside here, and to give us a few grapes for the table, if the children do not eat them all first. But in the past, some of the mountain *campesinos* would grow sufficient grapes to make all their own wine, *sí*. *Hombre*, making the wine was the most joyful task of the year for them – apart from killing the pig, *naturalmente*.'

'*Naturalmente*.'

'And when you have wine, you can have brandy also. All you need is a still, no?'

'You mean they had illicit stills up here?'

'Illicit?' The carpenter raised his shoulders and turned the palms of his hands upwards. '*Señor*, a still is only illicit if the authorities can prove that it exists. Therefore, there have never been any illicit stills in these mountains. *Me entiende*?'

I smiled approvingly. 'And, ehm, I don't suppose there are any of those little Mallorcan moonshine factories surviving today, are there?'

Juan Juan made a coy inspection of his feet. '*Indubablemente no*. The authorities are much too clever for that, they say. But, eh, there is one old fellow who lives on a high *finca* with an unobstructed view of the approach road all the way up the mountain. He has been constantly pissed and stinking of brandy for as long as anyone can remember, and yet no one has known him to buy a bottle of liquor – ever. *Es muy misterioso, no*?'

'Very mysterious indeed, Juan. Very mysterious indeed. Perhaps you and I could try to solve that little mystery some day. It could be a lot of fun.'

I looked at my watch, and already over an hour had passed since Juan Juan had predicted that this sojourn would take only half that time.

'OK, I know you're a busy man,' I said resolutely, 'so we'd better take a look at this tractor of yours, hadn't we?'

Juan quickly snapped his fingers as a token expression of urgency, disappeared into the darkness of the shed and emerged a few moments later pushing the twin of old Jaume's tractor – a chunky little two-wheeled *Barbieri* as spotless and shiny as the day it came out of the factory.

'*Por favor señor,*' he puffed, 'if you would now oblige by helping me to pull out the *remolque* and the other implements . . .'

We trundled the little trailer into the daylight, and I could see right away that neither it nor the rotovator and reversible plough which it was carrying had seen much service on the land. The metal parts, paintwork and tyres of the entire outfit were in mint condition, with not a smudge of dirt or trace of dust to sully the immaculate red-and-white livery.

'What do you think, *señor*?' gasped Juan Juan, breathless with exertion and anticipation. '*Es muy bonìito, mi tractor, sí?*'

I strolled slowly round the neat little display of equipment, studying every item from all angles and trying not to let Juan see how impressed I was. He was as nervous as a butcher's thumb, shuffling about behind me and eagerly praising every last detail of the gear-change system, the power take-off facility, the flywheel starting method, the easy coupling-up of the implements, the power of the diesel engine . . .

'OK, Juan,' I interrupted, not wishing to prolong his agony any longer, 'I like your tractor very much. It could be just what I need, but there is one very important detail . . .'

His face fell and he started to stammer anxiously, his hand on his heart. 'Oh, I promise you, *señor*, this tractor is *perfecto*. I bought it new only three months ago. Look, I have the receipt here. I have only used it a couple of times. It is *absolutamente perfecto*, and I am only selling it because I have discovered that, on the steep land up here, I would be better with a four-wheel-drive machine. It would have more grip and eh . . . and eh . . .'

'And a seat?' I suggested dryly.

Juan Juan wasn't sure how to take that comment at first, but when he realised that I knew full well that the main reason for his desired switch to a four-wheeled tractor was laziness, he gave me another of his playful digs on the chest and laughed with all the relief of a schoolboy who had just been excused for releasing a ferret up the art mistress's skirt.

'It's all right, Juan,' I grinned, 'I do want to buy your tractor, but the one important detail I need to know is the price. That's all.'

'Ah, *el precio*! I believe that this will be no problem for two gentlemen to agree upon.' He smiled hopefully, unfolding the receipt with slightly trembling fingers. 'You can see here the price I paid for *el tractor y los accesorios. Mira!* Beside it I have written the sum for which I am willing to sell. It is *exactamente* two-thirds of the new price.' Juan nodded his head gravely. *'Es un precio muy justo, no?'*

Still being unable to evaluate anything properly in pesetas, I did a lightning mental conversion into pounds sterling and calculated that I was saving over two thousand pounds on the showroom price.

'A very fair price indeed,' I agreed swiftly, grabbing the carpenter's hand to close the deal without further ado. This was the type of man I liked to do business with.

'Aha! Està bien! Està bien!' beamed Juan Juan, shaking my hand vigorously and patting me on the cheek, his eyes

sparkling with the luxurious prospect of owning a tractor with four wheels and, *mas importante*, a seat.

With instant and willing disregard for any other business which might have been awaiting his attention, *el carpintero* insisted on driving the tractor and trailer all the way down to Ca's Mayoral there and then, if I would be so kind as to go on ahead in his van and wait for him there, *por favor*.

To my surprise, our yard was a hive of activity when I drove in. The man from the sawmill was there with his mule and a cartload of logs, which he and the boys were busy stacking inside the *almacén*. Ellie was standing shouting *'Qué?'* at the foot of a ladder propped against the house wall, with old Pep wobbling about on top clutching a small sack in his hand.

'Oh, thank God you're back,' called Ellie when I stepped out of the van. She was clearly fairly perturbed about something. 'Maybe you can understand what this old nutcase wants. He came over just after the log man arrived and gave me a long lecture – something about *fuego* and *desastre*. He kept looking up at the roof. I couldn't understand any of it. And the boys weren't any help either. As soon as they sensed problems in Spanish, they were away and unloading those logs like a pair of beavers on piecework. I've never seen them volunteer for a job so smartly in my life.'

'All right, dear. Calm down. I'll try to find out what the hell's going on. *Hola, Pep! Buenos días. Qué es el problema?*'

'El problema, amigo, es la chimenea,' shouted Pep, letting go of the ladder and pointing wildly at the chimney as he swung about precariously on his lofty perch.

'What's wrong with it?' I yelled back.

'It is about to go on fire, that is what is wrong with it. *Coño*, if it is not seen to now, *inmediatamente*, you are going to have a *desastre*, a *catástrofe*. You did not know this?'

'No, I thought everything was –'

'Everything, nothing – somebody, nobody – somewhere, nowhere – *macarrones*, *cojones!*' snapped Pep. 'Never mind all that *basura*. I tell you, if I had not looked in this direction last night when I was tending my sheep on the other side of the valley – *coño*, there were sparks shooting out of your *chimenea* like moon rockets.'

'Right. So you're saying that the chimney needs to be swept before we light the fire again. Is that what you're saying?'

Pep fully extended both of his arms sideways and gazed skywards, adopting the unsecured crucified position some twenty feet off the ground while he beseeched the gods in *mallorquín* to inflict something which I didn't understand on *el loco extranjero* – me. Our old neighbour was in no mood to be trifled with.

'*Correcto,*' he growled eventually. '*La chimenea* must be swept and I have come to do it for you, but your *esposa* will not do what I ask. All she does is stand down there shouting "*Qué?*" while I flap about up here like a frog with his arse speared on the top of a bamboo cane. *Caramba!*'

'OK, Pep. I'm sorry about that. She just doesn't understand too well, you see. But if you explain what you need, I'll do my best to help. OK?'

Pep took a deep breath and spelled out his requirements with exaggerated calm. He would need me to get some old sheets, he bawled, and hang them round the mantelpiece in the inglenook. It was *muy importante* that I left no gaps and I should make certain that the sheets were securely weighted down on the hearth, because there was going to be a lot of soot coming down, and if it escaped, it would cling to everything in the house. *Seguramente*. *Ahora bien*, if I could do that now without further delay, he could get on with the job, then get back to more important work of his own.

'But what about the rest of your equipment?' I asked. 'You know, your brushes and ropes and weights and things. If you tell me where to find them, I'll bring them up to you.'

Pep muttered what I took to be a string of *mallorquín* swear-words, then snarled, 'Just organise the damned sheets in the house like I told you to do. I have all the equipment I need in this bag. *Comprende?*'

Ellie scurried after me as I strode off into the house to do his bidding.

'What was all that about?' she panted. 'He's in a real foul temper, isn't he?'

'Yes, but he's got our best interests at heart. He's insisting on sweeping the chimney right away, because if he doesn't, we risk a fire – a disaster, a catastrophe.'

'Yet another job that Señor Ferrer has neglected then?'

'Very likely. But let's just get some sheets draped round the fireplace before old Pep bursts a blood vessel up there.'

Ellie helped me do the necessary, then I hurried back outside to inform Pep that we were all set.

'Ready inside, Pep,' I shouted.

I looked on in trepidation as Pep clambered up off the ladder and scaled the pitched roof without any sign of fear for the certain fate which would befall him should he lose his footing on the slippery tiles. He reached the apex of the roof and flung one leg over the ridge, settling into a sitting position in front of the chimney stack. He then carefully fitted the open end of his sack over the top of the chimney and roared down at me, *'Bueno*, get back inside the house. And keep those sheets closed. Here we go! *Cuidado abajo!'*

I dashed back into the kitchen as Pep appeared to tip the contents of his sack down the chimney.

'Look out, Ellie,' I yelled. 'Watch out for the soot. It's on its way down.'

We heard a muffled screeching somewhere high up inside the chimney, followed by the raucous rasp of Pep's *mallorquín* cursing.

'Must be having a bit of trouble with his chimney-sweeping gear,' I said as placidly as I could to a highly-tensed Ellie.

A terrible commotion ensued. More wild screeching, more violent swearing, banging, clattering, scraping, thumping.

'What the blazes is that old loony up to?' gasped Ellie, staring wide-eyed at the ceiling. 'He's going to bring the whole house down on our heads!'

Just then, the first wads of soot began to thud down onto the hearth behind the sheets.

'It's all right now,' I said, patting Ellie's hand. 'Pep's got his equipment going at last. I – I just hope these sheets are secure.'

A frenzied beating and scratching continued to issue from the heights of the chimney, growing louder as Pep's sweeping apparatus worked its way downwards, pushing what sounded like tons of soot and loose masonry ahead of it.

'Must be some kind of mechanical device he's using,' I deduced nervously. 'Crafty old devil.'

Ellie was speechless.

Pep's machine screamed and clattered, the flailing rhythm of its mysterious sweeping components slowing, and the shrill noise of its motor deadening into a dull 'clack, clack, clack' as it finally settled into the deep heap of soot which we envisaged had now accumulated in the fireplace. Then silence. Behind the sheets, all was still. The job was done.

Ellie and I looked at each other, kneeling there semi-shocked and dumbstruck. We heard the sound of running feet outside the house, then the door flew open and old Pep burst in, wheezing desperately, sparks spitting from the mandatory cigarette wedged in the corner of his mouth.

'*Rápidamente!*' he spluttered, diving into the inglenook and hauling the sheets aside. 'Get the fucking thing out of there before it goes *loco* again and scatters all the damned soot. *Vámanos!*'

The sun shining through the open door cut a swathe of light through the dark cloud which was still billowing above the hearth. Thus illuminated, Pep fell to his knees and began to grope about wildly in the black pile.

'Got you, you dirty bastard,' he growled, grabbing his chimney-sweeper in both hands and fiercely shaking the soot from it. *Fantasma de puta!* You will never make a fool of Pep like that again, you –'

Pep's tirade of abuse was cut short by Ellie letting rip with a blood-curdling scream.

'Look!' she squealed in horror. 'That thing in his hands! It moved!'

I squinted closely at the nondescript black bundle, and sure enough, it did move – ever so slightly, but it did move.

Then a tiny eye blinked out of the soot at me. The thing was alive!

'Jesus Christ,' I whispered. 'I don't want to believe what I'm seeing. But it's true. The old reprobate has swept the chimney with a –'

'Wait a minute,' interrupted Ellie, steeling herself to take a closer look. 'No, it surely can't be –'

'Oh yes it is.'

'You don't mean –'

'Yes. The "clack, clack, clack" we heard a minute ago was –'

'Not . . . "cluck, cluck, cluck"?'

'Yes – the old bugger has swept the chimney with a hen! Ellie, he stuffed a live chicken right down our bloody flue!'

Ellie was clearly appalled, and her look of dismay did not go un-noticed by old Pep, who was now exposing his tombstone line of brown teeth in a self-satisfied grin while he shoved his befuddled bird back into the sack.

'Do not concern yourself, *duquesa*,' he croaked at Ellie. 'She was an unwilling sweep, this one. Tried to escape back out of the top of the chimney. It made things difficult for me up there, but I have to be honest – that kind of spirit in a hen always produces the best results. It is the extra flapping and clawing which does the trick. *Palabra de honor*.' He crossed his heart.

'No, it's not that at all,' I protested calmly. 'We don't doubt that a good chimney-sweeping job's been done, but my wife's worried about your hen, that's all.'

Pep raised an open hand and swung it down contemptuously. '*Mierda!* Do not concern yourself about the hen. She was off the lay anyway. Well past it, the old whore. *Puta vieja!*'

238

Ellie gasped at him in disbelief.

'No, I don't think you understand, Pep,' I went on. 'My wife here is only concerned about the condition of the hen, not if you're going to get any more eggs out of her. I mean, she must be badly knocked about after that trip down the chimney.'

'Bah, that is the typical confused thinking of a woman,' scoffed Pep. 'Perhaps the hen is a little bruised, but so what? A day simmering in the stock pot will soon fix that, no? *Va bé!*'

'He can't – he doesn't mean that, after all she's been through, he intends to wring that poor creature's neck now, and then . . . and then use her to make . . . soup?' hissed Ellie through clenched teeth.

Realising that his performance was getting Ellie well and truly riled, Pep dangled his sack provocatively in front of her face and bragged, 'One silly hen more or less is nothing to me. I have a randy old cockerel over there and he is always making new chickens for me anyway. He enjoys his work and he is good at it. *Es igual.*'

Pep heaved his lungs into a whistling, grating guffaw, which graduated involuntarily into an eye-watering fit of purple-faced coughing. Amazingly, the lit cigarette remained firmly ensconced in the corner of his mouth throughout.

I knew that this startling little bout of pulmonary convulsions had temporarily saved old Pep from a real roasting from Ellie, but mercifully, the futile conflict which would have ensued was averted by Charlie's timely entrance into the kitchen.

'The log man wants his *dinero*,' he announced, 'and another little guy has just arrived on some kind of horseless carriage contraption. You better come quick, Dad. The horseless carriage bloke doesn't look too well – just standing there shaking and muttering *"coñac"* over and over again.'

Poor Juan Juan was in a proper state of shock – beads of sweat standing out on his forehead, eyes staring, legs trembling, and his face a perfect colour match for the delicate yellow flowers which cascaded over the nearby mimosa tree in luxuriant bunches of tiny cotton wool globes.

'Are you sick, Juan?' I enquired anxiously. 'Can I get you anything? A glass of water perhaps?'

'*No, no, señor,*' he stammered. '*No aqua, gracias. Coñac – solamente coñac, por favor. Coñac! Pronto!*'

Having pre-assessed the emergency nicely, Charlie was already on hand with a bottle of Fundador brandy and two Churchill-sized balloons. I poured a goodly medicinal measure and closed the little carpenter's quivering fingers round the glass.

'Get that down you, Juan. It'll do you good,' I advised in a caring tone.

It was all the little fellow could do to get the glass to his lips without trembling the healing contents out onto the ground. Seeing Juan in such distress caused me to come over a bit wobbly myself, so I made instant personal use of the second glass that young Charlie had so thoughtfully handed me.

'Cheers, Juan. Your very good health. *Salud!*'

'*Salud, señor, y gracias.*' Juan gulped his brandy from the quaking glass with a slurping noise that reminded me of a calf drinking milk out of a bucket. Inevitably, he spilt some of the precious liquid, lost forever as it dribbled down his chin in golden rivulets. The man was seriously traumatised. I poured him another curative dose.

We were soon joined by the log man – a dour-looking fat chap with wild, unkempt hair and shaggy eyebrows joined together to give the appearance of two black, hairy caterpillars fornicating on his forehead.

'I'm afraid Juan Juan is a little *enfermo*,' I advised him. 'Nothing too serious, I'm sure. He'll be all right in a minute, so please don't worry.'

The black caterpillars knotted into a sinister embrace above his nose. He uttered a primaeval grunt and fixed his sunken eyes on the Fundador bottle. The state of Juan's health was evidently of no interest.

'Care for a brandy?' I asked superfluously.

The caterpillars danced and the log man grunted in the affirmative, greedily snatching my goblet from me.

'How about you, Pep?' I called to our neighbour as he ambled over from the house with his hen bag. 'Would you like a shot of Fundador to wash the soot out of your tubes?'

'*Basura*! Never touch alcohol. A water man, me. *Solamente agua*.'

'As you wish,' I said, somewhat surprised. For some reason, I had had old Pep pigeonholed as a master tippler, but the look of disgust on his face as he watched the log man devouring his brandy was patently genuine.

The log man dumped his empty glass unceremoniously on the ground, belched, arranged the caterpillars into a grim pelmet, then shuffled off in silence to his waiting mule.

'A real charmer, eh?' remarked Charlie. 'That guy was definitely standing in the wrong queue when the charisma was being dished out. I think his mule must have got his share.'

'Well, maybe he's just having an off day,' I said.

'Yeah, right off. He never even thanked Mum when she paid him for the logs. He counted every single peseta, but not a word of thanks. I thought he might have lobbed Sandy and me a few notes for helping as well, but nothing. Charmless berk.'

'Listen, Charlie, we have to face it – some people are going to be more difficult to break the ice with than others. Don't forget, we're the strangers here, so give the man the benefit of the doubt. Just assume he wasn't feeling too good today. Try to see the best in people. That's the thing to do.'

Charlie glowered disapprovingly and wandered off to join Sandy, who was examining Juan's tractor rig with an expression of stunned incredulity. 'Dad must have gone off his trolley if he's going to buy this off-road invalid chair instead of a real tractor,' I heard him mutter to his younger brother.

I feigned indifference and turned my attention once more to the little *carpintero*, who was now reassuming his natural human colour, thanks to the second large brandy, the dregs of which were now disappearing down his gullet. Not a spot of spillage now, I noted. A good sign indeed.

'That was a nasty turn you had there, Juan,' I commented. 'Maybe something you ate?'

'No food has yet passed my lips today, *gracias a Dios*. Otherwise I would have surely thrown it up again on my journey down from the mountain.' Juan paled slightly at the very thought.

I administered another good nip of brandy just in case a relapse was developing.

'Ah, *muchísimas gracias, señor,*' whimpered Juan, holding forth his glass with only the merest trace of tremor now. The treatment was working.

Old Pep lit another cigarette, spat squarely into the mimosa tree and settled back against the side of Juan's van to await the elucidation of the carpenter's problem which the triple prescription of Fundador was surely about to inspire.

'*Madre mía,*' moaned Juan Juan, clasping one hand a mite over-dramatically to his forehead as he sat down on the drawbar of the trailer. 'I have never been so scared. You see, I have not driven a tractor and trailer like this on the open road before today. And these two-wheeled tractors, *hombre*, they have no brakes – just a clutch, throttle and gears; so, to stop, you have to stand on the foot brake down beside your seat on the trailer. Without practise, it is a technique *muy difícil.*'

'That is the problem with a tractor,' interjected Pep scornfully. 'Does not understand the word "woah"!'

The carpenter paid no heed. 'It was bad enough coming down the mountain track from the *finca,*' he declared. 'I just stayed in first gear to keep it going nice and slow over the rocks and potholes. *Sí,* that was bad enough. *Bueno,* I was nearly shaken to death and I almost capsized a few times, but at least I felt in control . . . almost. But when I reached the main road, I tried to change into second gear, only to go a little faster, *poco a poco*, until I became accustomed to the smooth conditions. I had decided to be very cautious, to get the feel of everything before I encountered all those sharp bends.' His brandy glass rose to his lips at the memory.

'*Caga de toro!*' muttered Pep.

'It was then,' quaked Juan, 'that I made the fateful error, the mistake which lost me control of my nerves and almost cost me my life.'

'*Cómo?*' Pep enquired, inclining his head backwards and squinting sceptically at *el carpintero* through one half-closed eye which glinted darkly from beneath the shady canopy of his black beret.

Juan Juan swallowed a mouthful of brandy, grimaced and shuddered violently. He was either suffering a spasm of delayed-action shock or his empty stomach was finally objecting to the relentless battering of neat Fundador to which it was being subjected.

'Ah, that is better,' he coughed. 'I think I am regaining control of myself . . . almost. *Casi casi.*'

'*Sí, sí, carpintero,*' yawned old Pep. 'Just get on with it. What was this fateful error that you made? *Venga!*'

'You will not believe it, *amigos*,' Juan Juan confided solemnly, 'but when I tried to engage second gear, I missed the change and slipped the damned thing into top. *Madre de Dios*, when I let out the clutch lever, the tractor was off quicker than – quicker than –'

'Than a Jew's foreskin?' drawled Pep.

'Even more quickly than that,' enthused Juan, warming to his tale and quite oblivious to Pep's sarcasm. '*Hombre*, I shot down that road like a bullet from a gun. And – and I could not slow down. No synchromesh, so it was not possible to change to a lower gear, and with the weight of the implements on the trailer pushing us down the steep slope, the footbrake was useless also . . . even if I could have found it.'

His face started to twitch and his brandy hand developed a noticeable tremble again – but all to no avail. The forward-looking Charlie had already returned the Fundador to the safety of the kitchen. Juan Juan's self-control would have to proceed unaided, it seemed.

'*Dios mío!*' he continued, his voice yodelling with emotion. 'Those hairpin bends with nothing between the edge of the road and the bottom of the valley but five hundred metres of fresh air. *Jesús, Maria y José!*' he exclaimed, smartly invoking the support of a few holy spirits as a contingency back-up for those in the sadly-departed bottle. 'I thought the time had come to meet my maker. And when I encountered that coach coming up towards me on my side of the road at the switch-back bend above the Villa Tramuntana, I just closed my eyes, heaved the handlebars to one side with superhuman strength and prayed. *Hombres*, I could hear the panic-stricken *turistas* on the coach screaming as I hurtled past at breakneck speed.'

'*Incrédulo,*' croaked Pep, unimpressed.

'And when I opened my eyes – hey, the left wheels of the tractor and trailer were off the road, *en aire libre!* I saw people in the garden of the Villa Tramuntana below me running for cover. If I had not been going so fast round that bend, I swear I would have gone over the edge. I would have been done for. *Muerto!* I was saved only by the power of centrifugal force . . . ehm-eh, *y los santos*.' Juan nodded heavenward.

'So how did you eventually manage to stop?' I asked, offering my own silent prayer to Juan's saints for a speedy conclusion to this epic.

'Oh no, it was not possible to stop, *amigo*. Gravity pulled me faster and faster down the road – every corner a nightmare. I had to summon up hidden reserves of strength to wrestle with the steering round all those hairpins. I must have said a hundred *Ave Marias*. Believe me, it was only when I reached the level stretch of lane outside your gate that the machine slowed down and I was able to bring it to a stop. And eh, it was, of course, a miracle that I had even managed to turn off the main road into the lane in the first place. I was charging downwards at such fantastic velocity at the bottom of the mountain road there that I truly thought that I would shoot right past the end of the lane and on into Andratx. Can you imagine the *desastre* that would have resulted if I had burst onto the streets of the town on a runaway tractor and trailer? *Santo Padre!* Carnage . . . bloodshed . . . panic . . . mass death –'

'Mass crap,' Pep rasped through a shower of sparks from his *cigarrillo*. 'I watched you coming all the way down the mountain when I was on the roof. I had a perfect view from up there, and you were going so slowly that even the grandmother of my old mule could have trotted past you without breaking sweat. Your story is rubbish. *Basura!* Anything for a free brandy, eh?'

'Rubbish? What do you mean rubbish?' spluttered Juan Juan. 'It is you who are talking rubbish. No mule on earth could have matched the speed of my tractor coming down such a gradient. Everything was completely out of control and –'

'*Sí, exacto!* Completely out of control! And that would not have happened with the sure feet of a mule. Give me a mule any time. You can keep your tractors.' Old Pep was in full flight now. '*Caramba!* A mule will work for you all day long and will then present you with a free heap of shit to spread on your land, but at the end of the day, a tractor only leaves the country deeper in debt to some damned oil sheik.'

Juan Juan drew himself up to his full height and looked Pep boldly in the eye from a discreet distance. 'Do not tell me about heaps of mule shit,' he shouted. 'Who would need it? If you had been fighting a rogue tractor all the way down that mountain like I just did, there would have been more shit in the legs of your *pantalones* than a mule could pass in a year. *Vate a la mierda!*'

Having only just regained control of his nerves, the mild-mannered carpenter was now rapidly losing control of his temper.

Pep was delighted. First he had succeeded in winding up Ellie on the sensitive subject of poultry rights, and now he had Juan Juan going nicely on the perennial and insoluble rural argument of mules and donkeys versus tractors. He was having a wonderful time.

Fortunately, Ellie arrived back in the yard before Pep could stir up more strife. She was brandishing a packet of chicken-stock cubes.

'Here – take these with my compliments,' she said in English, thrusting the packet into Pep's hand. 'But please, spare the hen. These cubes will give you more soup than she would anyway. *And* you won't have to pluck them. So just let the poor hen go.'

Pep looked at the packet of stock cubes and smiled mischievously. '*Duquesa,*' he said, already quite aware that Ellie was the type of woman to be irritated by being addressed as 'duchess', 'I would not poison myself with these. Full of chemicals, like all your modern convenience foods. I do not put chemicals on my land and I do not put chemicals in my body.' He paused to think for a few seconds, then his face wrinkled into an ominous smile, heralding the delivery of the final riposte. '*Además*, I do not put chemicals into my *animales* either. They are fortunate creatures to have a kind master like me, no? *Por ejemplo*, when my dog has worms, I do not ask *el veterinario* to pump him full of any of those *tabletas sintéticas*. No, no – none of that synthetic muck for my dog. I use the ancient remedy provided by nature. I put a whole clove of garlic in his food every day. No stronger *antiséptico* than garlic, and it will flush the worms out of a dog even better than a hen sweeps a chimney.'

Ellie scowled at him in bewilderment. Although she didn't understand most of what Pep was saying, she instinctively disliked all of it, whatever it was.

'*Ah sí*, I can detect confusion in your eyes, *señora*,' observed Pep, adopting a suspiciously benign tone. 'You are wondering how I administer the garlic if the dog does not like the taste. What do I do if the dog eats all his food but leaves the garlic clove in his dish, eh? *No problema*. I do what all the smart doctors do with their fancy modern pills. I take the garlic clove so . . .' Pep placed the imaginary garlic on the tip of the upward-pointing middle finger of his right hand, 'and I stuff it straight up the dog's hole. *Puf!*' He jabbed the administering digit suggestively in the air and whispered with a serene smile, '*Un supositorio, sí?*'

The reddening of Ellie's cheeks indicated that she had probably grasped the gist of Pep's epilogue, and the undisguised titters from Sandy and Charlie registered their effortless comprehension of at least the obviously vulgar elements of the Spanish language. Juan Juan, conspicuously embarrassed, reverted to the study of his shuffling feet. I coughed unconvincingly.

Old Pep could not have been happier. His matchless repartee had overwhelmed the gathered company, in one way or another, so he could now afford to make some benevolent gesture appropriate to his proven position as top banana.

He straightened himself and held up his sack. 'Señora, I am grateful to you for the offer,' he said, touching the front of his beret in a show of new-found courtesy. 'Your chicken cubes would have been a fair trade for this old bird, but as I have explained, I have no use for such things. Nunca.' Pep handed the packet back to Ellie. 'But at the same time,' he continued, 'I am a compassionate man and a good neighbour, and as I can see that you are a sound judge of a hen, I wish to present this one to you. Here, the hen is yours. And you can tell everyone that the first livestock on your finca mallorquín was donated by me – Pep, un caballero generoso.'

Ellie was left standing dumbfounded, clutching the clucking sack, while old Pep swaggered off magnanimously towards his own finca to get on with more important things. He turned at the gate momentarily, however, and wafted the beret from his head in a low, sweeping bow. He was making the exit of a true gentleman.

'Viejo bastardo,' grumbled Juan Juan under his breath. 'Every time I meet that old goat, he makes a point of taunting me, and I always take the bait. It never fails. Es loco.'

'Yes, he certainly seems to delight in rubbing people up the wrong way,' I agreed. 'Yet beneath that gruff shell, he really does have a kindly side to his nature, don't you think?'

Juan nodded reluctantly. '*Sí*, I have to admit it. All the same, Pep can be hard when he wants to be – *muy duro*. Just try doing business with him. If he found two peseta coins in his change and one was more worn than the other, he would go to the bank and make them change the thin one for a new one. But he is also a benevolent man, and that is the truth. If that worn peseta was the only *dinero* he had in the world, he would gladly give it to you if you needed it. *Sí, Pep es un hombre muy benévolo*.'

He certainly was a contrary old character, I pondered, casting my mind back to the episode of the logs at Christmas. Without doubt, it would be fascinating to find out more about Pep and the story of his life. But that would have to wait until another time. For now, the most pressing item was to conclude the business with the carpenter that had started with a half-hour trip up the mountain almost three hours earlier.

I came straight to the point. 'Juan, I've kept you away from your workshop for too long. I must pay you for the tractor now without taking up any more of your valuable time. Will a cheque do?'

Sí, sí, un cheque, that would be fine, he assured me with a cavalier wave of his hand. But I had to be sure to make it out '*Al Portador*' and not to him by name. Making it payable to 'The Bearer' would keep the taxman off his trail, as I would understand. Greedier than an Osifar sludge pump, those tax-collecting *bandidos*. If they discovered that he had sold the tractor, they would only want to know where he got the money to buy it in the first place, he reckoned. *Hombre*, life was complicated enough without getting involved in all that income tax mess, so it was better to simply make the cheque payable '*Al Portador*'. *Sí*, that was the fairest system and one which served the business community well.

Juan Juan winked sagely, folded the cheque and secreted it carefully in a safe recess of his bulging wallet. No point in letting governments know too much about your finances, he confided, aiming another of his playful punches at my chest. Laughing heartily and enveloping us in the heady smell of Fundador, he shook hands with each of us in turn before settling in behind the wheel of his little van and driving out of the yard, cheerfully shouting *gracias* and *hasta la vista*, any thought of traffic police the furthest thing from his mind.

Nevertheless, my most vivid memory of that eventful late December day was to be the fiercely indignant sound of the hen's cackled protestations as Ellie assiduously washed the clinging soot from its feathers under the tap in the kitchen sink. The racket it made while being rammed down the chimney was but a twitter compared to that.

And what was Ellie's reward for her painstaking, humane efforts on the hen's behalf? When she finally tucked the squeaky-clean, blow-dried old bird into the snug hen's nest under the rickety chair by the well on the west terrace, the ungrateful fowl merely blinked in haughty incredulity, squirted a disdainful dropping in Ellie's direction, and took off as fast as its hen toes would carry it towards the gate, which it cleared determinedly in a cackling cacophony of flapping wings. It was last seen legging it through the twilight back to the comparative safety of old Pep's farmyard.

– EIGHT –

CITRUS SALES, TRACTOR TRIALS & A RODENT RECIPE

'Just look at that view of Andratx,' Ellie cooed dreamily. 'Every time I see it, I'm completely – I don't know – amazed. It's almost too spectacular to be true – like something out of an epic film, with the little houses all clustered below the mountains. It's – it's almost biblical, isn't it? I don't think I could ever tire of looking at that.'

We were returning home over the last rise on the main Palma road, having just driven the short distance along the coast to Peguera to introduce ourselves to old Jaume's friend, Señor Jeronimo the fruit merchant, who had promised without hesitation to come and *examinar* our oranges at the earliest opportunity. He knew our *finca* well, he had assured us with a ready smile, and *sí, sí, señores*, he was confident that he would be able to buy most of our crop. *No problemas*. He would only have to check the quality of the fruit and fix a mutually acceptable price, but that would be merely *una formalidad*, he was sure.

We could not have been more pleased. In fact, we felt on top of the world. Everything seemed to be falling neatly into place at last.

Even Sandy had recovered from his initial shock at seeing the tiny tractor which was to be our sole means of working the land, and when we had left for Peguera, he was busy attaching the plough to the little traction unit in preparation for making a start at turning over the weedy soil.

'May as well see how this Mickey Mouse machine works,' he'd grumped, surreptitiously polishing an almost imperceptible smudge of oil off the tractor's engine cowling with his cuff. 'Somebody has to do it. Bloody embarrassment.'

It appeared that Sandy had now decided to learn the essential basics of Spanish farming – in his own particular way.

Charlie, in the meantime, had volunteered to pick oranges ('for a modest fee') with Toni, his newly acquired friend from the village. As soon as they had reached the orchard, however, their baskets and secateurs had been abandoned on the ground while the two unlikely *labradores* disappeared upwards into the leafy depths of an old clementine tree, from where there soon emanated the broken-voiced, adolescent guffaws which greeted each swear-word and obscenity that they were swapping and keenly remembering in their opposing languages.

It appeared that Charlie had now decided to learn the essential basics of conversational Spanish – in *his* own particular way.

Ellie and I had felt glad and more than a little relieved to see that our sons were beginning to settle into their new lives, and this added to the welcome feeling of wellbeing which enveloped us during the return journey of our fruitful trip to Peguera.

'Care for a cup of coffee?' I asked.

'Don't ask stupid questions,' smiled Ellie, adjusting her sunglasses stylishly with one hand and giving me a pat on the knee with the other.

At the top of the Avenida Juan Carlos I, the long main thoroughfare which extends almost the entire length of Andratx town, is the Heladería Ca'n Toneta, appearing from the outside as just another of the street's less imposing shop fronts which the uninformed visitor could easily pass by without a second glance. It had become Ellie's favourite café in Andratx, not because its beverages were any better than those of any of the other bars in town, but more because, as she put it, there was something 'nice' about it. It wasn't just another smoky, newspaper-littered joint full of noisy, domino-playing males. It was peaceful, airy, spacious – in fact, usually quite empty. That morning was no exception – for a while.

We sat down in the deserted room and marvelled at our surroundings while we awaited the arrival of someone, anyone, to serve us. The Heladería, or ice cream parlour, had once been the town dance hall, according to Señor Bonet, the proud owner, and although the many discos which had opened up along the tourist *costas* had long since superseded the demand for such an *elegante* establishment as his, he was determined to preserve the interior in all its original 'ballroom' glory for the day when clientele of refinement and good taste would once more glide over his polished tile floor to the sweet sound of saxophone, violin, piano and drums.

While he patiently waited for the return of those halcyon days, he had closed off half of the dance hall with elaborate cane screens set between sturdy stanchions that supported the low beamed ceiling on several rows of decorative arches.

Globed chandeliers and vintage ceiling fans protruded into the room like exotic stalactites, and a row of old theatrical spotlights hung forlornly from a horizontal bar in front of the little curtained bandstand which was silent now and in darkness, save for the melancholy green glow from an illuminated sign indicating the adjacent gents' toilet.

Handsome, tapestry-upholstered oak chairs which would once have lined the opposite walls of the dance hall – the *muchachos* on one side eyeing up and being eyed up by the *muchachas* on the other – were now grouped round ranks of white plastic tables which, although functional, could hardly have looked more at odds with their baroque surroundings.

To one side of the main entrance, a pinball machine now stood at the ready with flashing lights beckoning aspiring wizards, while the floor immediately opposite resembled an exhibition stand at a trade show of electronic appliances for the bar trade. A perpetually-blinking gaming machine, programmed to shatter the silence every few minutes with short bursts of synthesised fairground music, stood shoulder-to-shoulder with a pair of tall, glass, refrigerator cabinets which were stuffed full of a bewildering variety of *La Menorquina* ice cream concoctions.

The bar, which occupied half of one side of the hall, appeared quite tidy and even slightly under-stocked at first glance, but closer scrutiny revealed that any spare space which might normally have been taken up with extra liquor bottles had been used to accommodate stands displaying potato crisps, packets of savoury nibbles oddly called 'BUM', tins of olives, sardines and baby clams, plastic cigarette lighters and even a selection of cheap watches. Pride of place among the bottles on the gantry had been awarded to a statuette of the Virgin Mary, although her position was now clearly under threat from a bizarre electric clock emblazoned with the hallowed badge and team picture of Real Madrid Football Club. Times were changing.

For the moment unmindful of such things, however, Ellie's stare was locked onto a glass cabinet at the end of the bar, and in particular to the mouth-watering array of sugary *ensaimadas* and huge fruit gateaux which adorned its top shelf.

'Oo-ooh, just run your eyes over that delicious cherry tart. Oh yes, I must have a wedge of that,' she drooled, just as Señor Bonet appeared from the back shop.

He was a cheery-faced man with a paunch to match, a shiny pate and a full set of black eyebrows to rival those of the log man, though altogether much more benign. In fact, Señor Bonet's caterpillars fairly bristled with bonhomie and good humour.

I ordered a *café cortado* for Ellie and a beer for myself. As was his wont, Señor Bonet afforded me the courtesy of asking which *marca* of beer I would prefer, and when I made my habitual enquiry as to which brands he had, his set reply was a polite, 'Estrella Dorada, *señor*.' He was sorry, but he only had Estrella Dorada today. The fact that Estrella Dorada was the only make of beer that he ever had in stock seemed of no real consequence to him. By inviting the customer to make a choice, he had at least gone through the mannerly motions which were to be expected of the *dueño* of such an *elegante* establishment as his, and that was the important thing for him.

'Could I have an *ensaimada*, *por favor*,' Ellie urged in her customary cocktail of English and Spanish, pointing hungrily at the cake display, 'and a large slice of the *tarta de*, ahem, cherries there. Yes, the *tarta de* cherries please. A *grande* piece.'

A look of helpless distress spread over Señor Bonet's face, his shoulders and eyebrows lifting simultaneously as if connected to a puppeteer's string. He slid open the back of the cake cabinet and solemnly tapped each item of confection in turn. They were all as hard as bricks.

'*Lo siento, señora,*' he lamented, 'but the *ensaimadas*, the *tartas* – they are all hard, much too hard to eat. *Son muy duras.*'

He looked mournfully at the cakes as if they were to blame for their solid state and for their continuing presence in the display cabinet so long after their sell-by date. These inanimate objects clearly had it in for Señor Bonet, and there seemed nothing that he could do about it. For reasons known only to himself, the obvious solution of removing the unsellable goodies and replacing them with a fresh selection didn't seem to enter into the scheme of things. He slid the hatch shut again, looked at Ellie through hurt eyes, and inclined his head to one side, his ear almost touching his slowly-raised shoulder.

Ellie joined in the sad shrugging and told him that she would just plump for a packet of 'BUM' then, *muchas gracias*.

No sooner had we returned to our table than a tiny, curly-tailed puppy came bounding out of the kitchen and made straight for Ellie. Like many town dogs in Mallorca, this one appeared to be a cross between a large Chihuahua and a small mongrel, and he was soon licking Ellie's hand excitedly as she bent down to tickle his chin.

'Ah-h-h, just look at his cute little pointed face,' she crooned.

Predictably, the little doggy licks quickly turned into playful nibbles which rapidly developed into serious attempts at hand-eating, so when the first blood was drawn from her fingers by the needle-sharp teeth, Ellie gave the pup a sharp rap on his cute little pointed nose, and made a firm offer to lodge the toe of her cute little pointed shoe under his cute little curly tail if he didn't get lost *pronto*. 'SCRAM!'

The pup got the message and took off briskly in search of a playmate without such a bad attitude towards fun and games. He didn't have far to go.

A rotund German matron in mandatory hiking gear had just drifted wide-eyed into the Heladería in the company of her myopic daughter, a gangly senior fraulein in her forties with thick spectacles, the beginnings of a respectable moustache on her top lip, and a fixed, confused grin on her face. The august Deutsch duo were drawn magnetically to the inviting array of cakes and sweet pastries in the glass showcase by the bar, and after a harmonious chorus of Teutonic 'Oo-ing', 'Ah-ing' and lip-smacking, the mother rattled out in expansive German what seemed to be an order for the entire display of *ensaimadas* and *tartas* – the lot, *alles*, *das ganz*.

Señor Bonet's pained expression and flawless routine of cake-tapping amid woolly apologies for the inexplicable presence of this inedible fare in his cabinet met with frowning silence from the frau and an increasingly confused grin from the fraulein.

'Warum?' boomed the incredulous matron at length. *Mein Gott!* Why keep stale *kuchen* in *das kabinett*, she demanded to know – especially when *die kuchen* still looked *so gut?* Was *mein host ein total idiot, ein kretin, ein dummkopf?*

Señor Bonet nodded innocently and gave the frau a genial smile. He evidently was not a student of the German tongue.

Fuming with frustration, the old frau hiker pressed a forefinger to her temple, pointed aggressively at Señor Bonet's head and bellowed, *'Kaputt! Kannst Du mich verstehen? KAPUTT!'*

He closed his eyes passively, gestured towards the row of cakes, and sighed, *'Sí, señora. Kaputt.'* All of his cakes, he agreed resignedly, were *absolutamente kaputt*.

Come what may in the way of provocation, no world war was about to be started with the collaboration of this gentle man.

The pup, who had been sitting unnoticed at the Germans' feet during their abortive cake-buying campaign, looked up in animated anticipation as their order was finally changed – after a storm of huffing and puffing from *die Mutter* – to two jumbo *bocadillo* sandwiches of long, crusty bread rolls filled with slices of best *serrano* ham. Now, this did look promising.

The two German ladies marched off, the pup trotting efficiently behind, and sat down with their Amazonian snacks at a table near the glass door which opened onto the back patio.

The pup chose that moment to announce his presence with a piping yelp and a mischievous tug at the laces of one of the fraulein's stout trekking boots.

'*Oo-ah-oo, Mutti,*' she giggled in a girlish treble that belied her distinctly butch demeanour, '*ein hund – ein junger hund! Ha-a-I-I-o-o! Ach, so klein, so schön!*'

She lowered a tentative hand to the leaping pup, who immediately sank his teeth into her thumb and held on grimly, growling as menacingly as his infant voice would permit – his ears clapped back, his eyes staring like a rabid fox. The fraulein's show of affection had been a sham, and the pup knew it. This human was shit-scared, and as things stood, that could only mean two things for the pup – fun and food. He was going to play this game for all it was worth.

'EE-EE-K!' screamed the fraulein in agony. '*Meine hand, Mutti. MUTTI . . . der hund! Gib' ihm etwas zu essen! Hilf'e! Schnell! SCHNELL!*'

The old frau's frown deepened, but she heeded her daughter's pleas nonetheless, flipping open the fraulein's *bocadillo*, tearing off a piece of ham and hurling it contemptuously to the floor.

The cunning ruse was an instant success. The pup released his grip on the thumb and transferred his attentions to the morsel of ham which he proceeded to gnaw at contentedly.

'*Ah-h-h, dankeschön, Mutti,*' whimpered the big fraulein. '*Vielen, vielendank.*'

Peace reigned as the pup enjoyed his little treat and as the two peckish *damen* settled down to tuck into their *bocadillos*. But the truce was brief. No sooner had the pup gulped down the piece of ham than he was at it again – snarling and yapping and scratching with his front paws at the skinny expanse of bare flesh which stretched from the bottom of the fraulein's moleskin knee-breeches to the top of her rolled-down socks.

'A-a-ee-ee!' she screeched, launching her heavy boots into a frenetic goose-step under the table. '*Hör auf! Bitte! BITTE!*'

But the pup had no intentions of obeying her pitiful appeals to stop – not until another lump of ham was forthcoming, anyway. The mumbling *Mutti* was forced to oblige, and another few moments of respite followed while the pup demolished his second helping of *serrano*.

This noisy pantomime was repeated every minute or so until the pup had devoured every last scrap of meat from the semi-hysterical fraulein's sandwich.

Señor Bonet, who was contriving not to notice any of this on-going farce, had already decided to mask the clamour of the minor Germano-Hispanic conflict by inserting a music cassette into the ghettoblaster which was rather impiously stationed on the gantry under the statuette of the Virgin Mary. So now the romantic strains of Julio Iglesias singing 'Begin the Beguine' were belting out in direct competition to the racket of combat coming from the Germans' table. Julio had a struggle on his hands.

By this time, the show was also being watched by two young kitchen maids who had slipped quietly out of the back room and were sniggering their way through an unexpectedly entertaining coffee break at a table immediately behind the Germans, and with feigned indifference by three local worthies who had installed themselves on stools at the bar, perhaps lured by the pandemonium to desert their habitual posts at the usually more lively Bar Cubana across the street. Two electricians who had also arrived on the scene were doing their best amid the bedlam to concentrate on repairing something inside the works of the stainless steel cold store.

Meanwhile, the pup, having succeeded in scoffing all of the *jungfrau's* ham, decided to continue his highly lucrative sport by diverting his attentions to the feet of her mother.

'There's no way he's going to get the better of that old battleaxe,' I said to Ellie. 'She'd frighten the daylights out of a pack of wolves, even without her monocle and spiked helmet.'

Sure enough, at the first tug on her bootlace, the old frau thundered, *'NEIN! DAS IST VERBOTEN!'* She snatched the startled pup off the floor, ripped half of the *serrano* out of her own *bocadillo*, stomped over to the glass door, kicked it open, and flung the pup and the ham outside with a triumphant shout of *'RAUSS! SPANISCH SCHWEINHUND!'*

Slamming the door shut, she brushed her hands together with resounding claps, and swaggered back to her table, *grossen brosten* thrusting purposefully forward beneath her heavily embroidered Bavarian cardigan. That was how to do it, she confidently informed her cringing daughter. *Hunden* were the same as *personen*. *Himmel!* You had to show them who was *der meister. Ja! Bestimmt!*

'I'm afraid you were right,' admitted Ellie. 'The poor little mutt bit off more than he could chew there.'

'No, no, *señora*,' one of the kitchen maids whispered to Ellie. *'Es OK. Mira!'* She pointed discreetly towards the patio, where the pup had already gobbled up all of the frau's half portion of *serrano* and was now scampering at top speed –

ears flapping and tail at the high port – round the little terrace and back inside the café through the open rear door.

The kitchen maids nodded in expectant glee as the self-satisfied old German bent over to retie her bootlace, her ample buttocks pointing towards the very door through which the galloping pup was now making his re-entrance.

'*Pass auf, Mutti!*' yelled the frantic fraulein. '*Der hund kommt! ACHTUNG!*'

But it was too late. The pup had already disappeared at full tilt up the old frau's Tyrolean dirndl, and his summary act of revenge was heralded by a blood-curdling scream and the clatter of dishes as *die Mutter's* flailing arms swept the *bocadillo* plates cleanly off the table.

The humbled Germans were obliged to run the gauntlet of derisory glances and smirks as they stumbled and bumped their way between the plastic tables, making a decidedly ignominious exit while the pup happily guzzled the remaining half of the frau's ham off the floor.

'*Todo va bien,*' grinned Señor Bonet, emerging from his diplomatic hiding place behind the bar. All was well indeed. As our landlord was delighted to point out, not only had the German *señoras* bought an expensive ham lunch for his pup, but they had also left behind the two uneaten *barras* of bread which would provide a tasty free peck for his hens. Who was the *dummkopf* now, he asked with a wry flick of the eyebrows?

The three worthies and two kitchen maids chuckled their approval of the entire episode. *Ah sí*, they chanted, who was the *dummkopf* now?

We were tempted to sit on and see what further *comédies sans frontières* might be propagated by Señor Bonet's cake cabinet, but like all good things, our short interlude of international café society Andratx-style had to come to an end. It was time for Ellie and me to get back home to find out what progress Sandy and Charlie had made with their respective chores.

We slipped out into the comparative quiet of the street, only to be immediately beckoned back to the Heladería's entrance by the waving proprietor.

Tomorrow night would be the last night of December, he reminded us, and he and Señora Bonet would be throwing a party to bring in the New Year. All of their regular customers would be there, and he hoped that we, as *clientes muy estimados*, would honour him with our presence. Our sons, our *hijos guapos*, would also be most welcome to join in the festivities. *Hombre*, it was going to be *una fiesta espectacular*.

'*Muchísimas gracias,* Señor Bonet,' we replied without recourse to debate. *Hombre*, we could hardly wait.

Roars of '*Bravo!*' and '*Olé!*' rang out from inside the high stone wall as we drove down the lane towards Ca's Mayoral, prompting us to wonder if our sons had taken up bullfighting lessons during our absence. But we needn't have worried. The cheers were coming from a small group of our elderly male neighbours who had assembled by our *casita de aperos*, an ancient stone implement shed of miniature proportions, to watch nothing more dangerous than Sandy's first attempt at ploughing with the little *Barbieri* tractor. On reaching either headland of the field, he would nimbly swing the machine through one hundred-and-eighty degrees, while deftly pulling a lever to trip the reversible plough in readiness for the return draught over the land. Then up would go the shouts of his audience, the volume of the acclamations depending on their opinion of the comparative straightness of his last furrow.

We could hear a raucous outflow of instructions and curses which suggested that old Pep was foreman of the jury, and judging by the permanent grin on Sandy's face and the good-humoured banter flowing from his audience, our boy must have been doing all right. We thought it best to let him get on with his self-motivated process of integration without interference from us, so we slipped unseen into the house.

To our astonishment, Charlie and Toni had picked a fair quantity of oranges, which had been arranged neatly into four groups – according to variety – on the *almacén* floor.

'It's most unlike Charlie to leave a tidy job like that,' Ellie said sceptically.

'Yes,' I agreed, stroking my chin, 'it's all very suspicious, and Charlie doesn't even know one variety of oranges from the other. This is weird.'

'Don't get fazed,' said Charlie dismissively, breezing into the storeroom. 'None of that was my idea. This guy – ehm, Sitting Bull or Cochise or something –'

'I don't suppose you mean Jeronimo, by any chance?' asked Ellie.

'Yeah, that's him, Mum. Dead right – Jeronimo. Great name. Anyway, he came down to the orchard when you were away. I couldn't understand much of what he was saying, but Toni knows a bit of English from school, so we managed to have a pow-wow . . . get it?'

Ellie and I stared at him blankly.

'OK, suit yourselves. Anyway, the good news is he's going to buy as many of your oranges as he can. He says they're OK . . . considering the hard up state of your trees, but he can only pay you about seventy-five pesetas a kilo wholesale, so he reckons you should try some of the local shops too. You should be able to get a better price from them, he said.'

'So what's the reason for the four little piles of oranges here?' I asked.

'That was Jeronimo's idea. See how each orange has got one or two leaves left on the stalk? Well, you have to do that to show that they're fresh Mallorcan oranges. It's a good tip. Very important, he said. Of course, it means that you have to sell the oranges before the leaves wither. That's the whole point. Anyway, he came in and divided the oranges into these varieties for you, so that you can show them off to the shops properly. No point in doing it any other way, he said.'

I raised a pleasantly surprised eyebrow. 'Well, well, Charlie, for someone who can't speak the language very well yet, you've managed to absorb a lot of handy information this morning, haven't you?'

'Yeah, yeah, but we have our ways, Dad. Foreign languages pose no barrier for the smart modern kid.'

'Oh yes? Well that's something which will be put to the test when you start school out here next week, because Spanish is compulsory.'

'No sweat, *padre*. I've already picked up a few useful Spanish phrases from Toni. Oh yeah, you can bet that the Spanish teacher is going to be well impressed by this new boy's lingo.'

Ellie grabbed him by the scruff of the neck. 'Listen to me, young Einstein. If you mean the choice items of vocabulary which I suspect Toni was teaching you up the clementine tree, I would suggest that you don't repeat *any* of them at school, or you won't be a new boy there for long. You'll be expelled on the first day.'

'Now then, Mother dear, that *is* a brilliant thought,' replied Charlie with a wicked glint in his eye.

'Adiós, Pep. *Adiós,* Jaume, Bartólome, Francisco. *Gracias. Adiós.'* It was Sandy shouting his farewells to his venerable mentors, backing into the house and bowing in acknowledgement of their affable plaudits, which we could hear being yelled out as our old *vecinos* wandered off into the lane.

'So how did the ploughing go?' I asked. 'If the enthusiastic noises coming from your old pals were anything to go by, you must have been making a pretty good job.'

'Uh-huh, it was OK . . . I suppose,' said Sandy, trying without success to disguise the obvious pleasure which he had derived from his morning's work, 'but not what we're used to. It takes some time to get the knack of operating that dinky tractor and plough – walking behind the machinery instead of sitting on top of it. It's a bit like a mechanical horse, but once you get the hang of it, it's not so bad. Mind you, I had plenty of advice from old Pep and his team of ploughing professors, so I couldn't go too far wrong.'

Just then, the said Pep, breathless after a short dash from the field gate, stuck his head round the *almacén* door. *'Perdón, amigos,'* he puffed, but when he had seen our car in the yard, he had felt compelled to come back to give us this *información importante*. This *chico* Sandy, this young *colega,* he was a natural with the plough – almost as good as Pep himself had been at that age. But if we wanted our son to perfect the technique, to become *un maestro* of the art of ploughing like him, we had to scrap that damned tractor. A mule, *amigos* – buy a mule *inmediatamente* for the *chico*. That was *esencial y urgente*. '*Hasta luego!*' He banged the door shut and was gone.

Sandy looked at me uneasily and raised his hands. 'Please, Dad – don't even think about it.'

'Take it easy, Sandy. *Tranquilo*. A mule isn't on our shopping list. No, I think a two-wheeled tractor with handlebars is as far back in time as we need go – no matter what old Pep says.'

I gave both boys a hearty slap on the back. 'Anyway, lads – thanks a lot. You've both done well this morning, and I really appreciate it. We both appreciate it, don't we, dear?'

'More than that,' beamed Ellie, planting unwanted kisses on their cheeks before they had a chance to take evasive action. 'It's wonderful to see you both getting involved in our new venture here, and enjoying it too. It makes me feel, well – nice. Listen, why don't we head off round some shops with these oranges, and if we time it right, we might just arrive somewhere good for a bite of lunch. I think we all deserve a treat. So come on, you two. Get cleaned up, and let's go!'

Despite our best efforts at salesmanship, old Jaume's predictions were to be proved all too correct. The response from every little shop we visited in Andratx was the same. Every single *colmado* and *mini-mercado* already had more than enough oranges being supplied by local farmers, so there was no need for them to buy any more, *muchas gracias*. Even a desperate ploy of bullying our two sons to go into the little stores and make smiling spiels in broken Spanish in a cheap attempt at charming the elderly lady shopkeepers proved futile. They simply did not need any more oranges, and that was that.

'Only one thing for it,' I concluded after we had been rebuffed by every potential customer in town, 'we'll have to cast a wider net. Let's head for the hills. We'll try a few of the mountain villages. There must be a keen demand for oranges up there, especially if we offer to make regular deliveries of fruit fresh from the trees, so to speak. What do you all think? A good idea?'

It was quite apparent from the silence that greeted my breezy suggestion that Ellie and the boys didn't share my optimism, but with the prospect of lunching out in the balance, they allowed their stomachs to rule their heads. They humoured me.

'OK, boss,' they droned. 'Let's head for the hills.'

We followed the hairpin route of Juan Juan's epic tractor trip – albeit in reverse direction – up the precipitous north-western slopes of the Sierra Garrafa, then on through the wooded high valleys, past isolated farms with inquisitive little flocks of shiny black goats, climbing steadily until we finally reached the giddy heights of the Sa Grua Pass. There, where the narrow, unfenced road clings impossibly to the sheer mountainside, Ellie and Charlie – who were sitting by the car's nearside windows – were treated to an uninterrupted, vertigo-inducing glimpse of the bottom of the gorge, so far below as to resemble the view from a slowly ascending aeroplane.

As we started the long descent into the next valley, the pretty village of Capdella could be seen far ahead, nestling peacefully in low hills between the massive bulk of Son Font and Sa Grua mountains, while below us and away to our right, the bed of a little *torrente* meandered through the rolling vale of Son Vich, its winding course etching a thin ribbon of green through the bare soil of the almond groves which fill the valley on its ever-widening path to the distant sea.

At the sight of these stunning views, the car was filled with gasps of admiration, ambiguously mingled with sighs of relief at having survived the hair-raising drive over the mountains – a journey which we might just as well never have made, for the purpose of advancement in the fruit trade, at any rate.

We met with the same negative replies at every *colmado* we called at in the sleepy little communities of Capdella and nearby Calvià. Even though there were few citrus orchards in the high land surrounding these villages, there were, nevertheless, sufficient supplies of oranges to satisfy local needs; more than enough, we were told repeatedly. *Demasiado, gracias.*

I finally parked the car in the tiny *plaza* opposite the Bar Bauzá in Calvià, and while the boys trudged off grudgingly with their trays of oranges, Ellie and I took time to watch the quiet, unhurried progress of a few townsfolk who gathered to fill their pitchers with the pure mountain water which trickled forth from the communal spring at the side of the square. It was a rural scene which had probably changed little in many centuries, save for the incongruous modern convenience of an aluminium telephone kiosk on the pavement, for the rattly passing of an occasional tradesman's van, or for the means of transport now favoured by the typical *campesina* wife in town for her shopping – a Mobylette autocycle in place of the donkey and cart. But the aura of timelessness was restored by the sight of two grey-clad nuns serenely climbing the adjacent flight of stone steps leading to the huge parish church, its lofty twin domes watching over and dominating the little houses of the town and every lowly *finca* for miles around.

After hours and no success, we decided to make the mountain village of Puigpunyent our last call. Our empty stomachs were beginning to make us irritable and I'd heard of a good little restaurant.

The drive to Puigpunyent was perhaps less dramatic than the first leg of our journey, and certainly more relaxed. We travelled through some of the most pleasant countryside to be seen in that part of Mallorca, leaving the gently sloping fields of almond and olive trees around Calvià, and passing remote, honey-stoned farmsteads standing in walled meadows with flocks of long-legged sheep foraging among the gnarled trunks of age-old carobs, their dark green canopies a permanent haven of shelter or shade for the grazing animals throughout the four seasons.

The wide upland valley gradually funnelled into a little canyon as we approached the lower slopes of the Sierra de Cans, our route twisting between densely wooded crags, the branches of trees on either side almost touching in places and filtering the sunlight into dappled patterns on the road ahead.

An ever-changing landscape greeted us round every curve of the road as it climbed and snaked through the woods which blanket the lower slopes of Bauzá and Cans mountains. Clumps of heather were already beginning to produce little pink flowers in the roadside clearings, where circles of charred stones and a few empty wine bottles marked the places where Palma families would stop on their weekend outings to cook *paellas* under the scented pines.

Descending at last to meet the main road which links Palma and Puigpunyent, I noticed that the pinewoods had become increasingly interspersed with evergreen oaks, their grey-green foliage lending a more mellow tone to the surroundings, an almost tamed quality that was in marked contrast to the rugged terrain through which we had just driven, and which became more noticeable the farther we travelled towards our destination. Even the road seemed smoother now – straighter and faster as it carried us west towards Mount Galatzo, whose magnificent summit towers above all others in that area. We sped past cherry orchards bounded by high hedges of cypress, the manicured fruit trees gaunt and naked in their winter slumber, while the bank of the *torrente* which flanked the left of the road was hidden under a verdant cover of myrtle and reeds, of blackthorn and cane, growing in profusion among ancient carob and olive trees and the now-ubiquitous evergreen oak.

By the time we reached Puigpunyent, the landscape had softened into a lushness more reminiscent of a forest glade in the South of England than of a remote valley high in the Mallorcan mountains. Tall elm, poplar and ash stood steadfastly together in a dell on the periphery of the village, their stout boughs and profuse network of upper branches paying perpetual homage to the supply of precious water carried to their roots down the rocky beds of the *torrentes* which spring from the storm-rain catchments of the surrounding mountains. To complement the natural beauty of this languid woodland nook, enlightened villagers of a bygone era had planted each side of the main street with elegant ranks of plane trees, mature now, with delicately camouflaged trunks of peeling bark blending subtly with the muted colours of the weathered walls and roofs of the old stone houses.

During the stifling summer months when the coasts and plains of Mallorca are baking in the relentless heat of the sun, Puigpunyent is an oasis of glorious shade provided by its leafy deciduous trees, and of refreshing, cool air wafted down on gentle breezes from the heights of Mount Galatzo. Even on that pleasantly warm winter's afternoon, there was a wonderful freshness about the environment there, both stimulating in the magnificence of the mountains and soothing in the feeling of quiet security exuded by the steep cobbled lanes of cosy, terraced cottages and the smell of smoke from their olive wood fires. It was an aspect of a rare face of Mallorca that time seemed hardly to have touched – but for the wheel-less, stripped shell of a Seat Panda car which had been conveniently left in the main street for use as a communal rubbish skip.

'Thank God it's closed,' yawned Charlie, looking out of the car at the shuttered windows of the *colmado*.

'Agreed,' said Sandy. 'The shopkeeper's had the good sense to go for lunch, and I suggest we do likewise. I'm weak with hunger.'

'You bet,' said Charlie, perking up at the thought of food. 'My stomach's so empty I think it's started to digest itself. It's cruel to keep a growing boy like me away from grub for so long. Look at the time. It's nearly half-past two!'

The Bar-Restaurante Es Pont was situated back at the cross-roads near the entrance to the village and close to the bridge from which it takes its name. Shade for Es Pont's big arched windows had been created in typical Puigpunyent style by the planting of young willow and pepper trees along the pavement outside, and although the building was of more modern construction than the majority of the village houses, it had a welcoming air, and the clutch of little vans parked nearby testified that this was indeed a good place to eat – and good value too, according to the blackboard by the entrance, which advertised a very reasonably priced three-course *Menú del Dia*, including bread and wine. There was no holding us back.

I could sense Ellie's feeling of disappointment, however, when we entered the large, Spartan room in which the only concessions to comfort appeared to be a few basic tables and chairs with a scattering of the day's newspapers, a wood-burning stove in the middle of the bare floor, a colour TV on a high shelf showing a dubbed-Spanish episode of *Neighbours*, and a pool table. But needs must, so I stepped resolutely to the bar, where several dusty building workers and blue-overalled delivery men were already enjoying their post-lunch coffees, liqueurs and cigars. Sandy and Charlie made straight for the pool table, all symptoms of starving to death miraculously relieved at the first glimpse of the blue baize.

The barman was a robust old fellow of taller than average height, his chubby, deadpan features accentuated by a flat cap which looked several sizes too small and was adorning the top of his head like a tiny flying pancake that had landed on a friendly melon. He was drying glasses.

On enquiring of him whether it might still be possible to order lunch, I was directed by a sideways dip of his head towards two wide archways at the rear of the room, one of which had been made impassable with a barrier of potted palms and some hanging flower baskets. Ellie and I stepped apprehensively through the other, the babble of male conversation hushing into stony silence while Ellie was appraised thoroughly by a dozen pairs of arrantly appreciative Latin eyes.

I spied a waitress emerging from the kitchen with a row of brimming plates balanced along the length of her arm, and I went off to ask her for a table, leaving the blushing Ellie with no refuge other than to simulate a deep interest in the smell of some artificial flowers which had been creatively arranged inside the propped-up lid of an old writing desk at the side of the archway. To her further chagrin, Ellie's coy smoke-screen tactic did nothing but encourage a low rumble of knowing chortles from her artisan admirers.

'I know it's sometimes an inevitable price to pay for having good lunches in these places where the workmen eat,' Ellie whispered once she was safely seated in my lee at a quiet window table, 'but I do wish they wouldn't be so blatant about their eyeing-up. It's so embarrassing.'

'It must be, and I do sympathise,' I said in a matter-of-fact way, 'but it doesn't look like changing that quickly around here, so maybe you'll just have to try to adopt a more liberal attitude.'

'Liberal attitude?'

'Yes – more, well, you know . . . Continental.'

'Continental! I'll tell you something for nothing – I bet none of the men in here would dare to indulge in their 'Continental' letching if their wives were with them,' retorted Ellie. 'And I'll tell you why. Because they'd get a liberal Continental kick up the liberal Continental wedding tackle, that's why!'

I accepted her prognosis without question.

'You know, it's a really nice restaurant, this,' I said, tactfully changing the subject and glancing round the bright, airy room. Marble-topped, wrought-iron tables were set out spaciously between the inner archways and two complete walls of arched windows which framed views of spreading plane trees and a covered terrace where more tables stood surrounded by earthenware tubs of climbing flowers. The comforting sound of smouldering logs crackled in a carved stone fireplace next to our table.

'You're right. Who'd have thought there'd be such an elegant eating place tucked away behind that smoky bar room,' she said, peeping over my shoulder at the remaining customers, most of whom were now giving their teeth a final pick and swilling down the last of their wine and *gasiosa*. 'It's more than just a cheap workaday canteen for those *muchachos*, by the looks of it.'

'Yeah – this'll be how they earn a coin during the week, but it's a cert that the better-off Palma lunch set will take over at weekends. Two restaurants in one, and the tables always full, eh?'

At that, a frail, old village woman, who had been quietly eating with her husband at a table by the kitchen door, stopped beside us on the way out and, first having begged our forgiveness for her intrusion, shakily plucked a fresh rose from a vase on the mantelpiece and handed it to Ellie with a shy smile. Speaking softly in *mallorquín* – with the occasional Spanish word thrown in by way of translation – she said to Ellie, a kindly twinkle lighting up her eyes, 'This flower has a much nicer perfume than the plastic ones in the old desk over there, no?'

Ellie was stuck for words.

The old woman then made a contemptuous gesture with her hands, as if swatting troublesome flies from the vicinity of her ears. 'You should pay no heed to these men who come in their vans, *señora*. They are city men – all of them originally from the mainland, of course. You can tell by their foreign accents. *Sí*, they are *españoles!* Mallorcan men do not treat ladies in such a *manera ignorante* – certainly not the men from Puigpunyent!'

She turned to glare at her elf-like husband who nodded his head obediently – while giving Ellie a mischievous wink unseen by his wife, and patting me lightly on the back. He took his wife's arm and they tottered off falteringly into the afternoon sunlight.

'What a nice gesture,' sniffed Ellie, dabbing her eyes with her napkin. 'What a lovely, sensitive old lady.'

'Not so lovely and sensitive towards the mainland Spaniards, though. Like a lot of old Mallorcans, no love lost there, it seems.'

'Got a cold coming on, Mum?' asked Sandy, the irresistible magnet of food smells having finally torn the boys away from the pool table.

'No, your mother's just made a new friend here, and she's a bit overcome by the experience,' I joked.

'Not that ancient bint who wobbled out with Mr Magoo just now?' grinned Charlie.

'Yes,' Ellie said, flicking a little tear from the corner of her eye, 'and she is *not* an ancient bint. She's a lady – a lovely old lady . . . really nice, so show some respect.'

'No pressure,' shrugged Charlie. 'It's just that she went up the street and unlocked the door to the *colmado*, that's all.'

'So you get the picture, Mum?' Sandy smirked. 'Your new friend might also happen to be the local shopkeeper. So go for it. Ride your luck. Hit her with a deal for our oranges. This could be the break we've been waiting for.'

'Hold on lads,' I said, shaking my head. 'Don't get me wrong – business is business, and your idea's a good one, but it's just not on.'

'Why? Where's the problem?' Sandy asked.

'It's just that Puigpunyent is a lot farther from Andratx than I had thought – too far to be worth the trip for the few kilos that we'd sell in a little village like this, even if they wanted our fruit.'

'Whew – that's a relief,' Ellie sighed. 'I wouldn't have relished making that journey on a regular basis anyway.'

'So we'll have to make do with what Jeronimo pays?' Charlie asked.

'Looks like it,' I replied.

'We aren't going to make much of a living from that – seventy-five pesetas a kilo – about fifteen pence a pound,' remarked Sandy. 'Not unless we live on a diet of oranges ourselves, that is.'

'OK – I think we've had enough orange troubles for one day,' Ellie said, rubbing her hands together. 'Let's eat. What's on the menu, Peter?'

'Right – it's either *Arroz Americana* or *Judías Blancas de la Casa* for a kick-off, followed by *Costelles de Porc* or a special called *Greixera de Rates Sa Pobla*. The special costs a bit extra, of course. Then, for dessert, would you believe, it's . . . *Naranjas de Andratx*.'

'Oranges from Andratx!' came the concerted cry of disbelief.

'That's right,' I sighed. 'Somebody's beaten us to it after all.'

'Ask them if they'll give us a discount if we bring our own oranges in from the car,' Sandy mumbled.

'Yeah, I'd even go and get them – for a modest fee,' volunteered his younger brother.

'Forget oranges,' I pleaded. 'Just make up your minds about what you want to order. Now, there's a rice dish of some kind or the house beans concoction up front, then it's either pork cutlets or the special – but I have to say I'm not too sure what the special is, OK?'

'But the special *is* more expensive?' checked Sandy.

'Yes, but I've a hunch you may not be too keen to –'

'Don't worry – if it's dearer, there's liable to be more food on the plate, and that's all I'm interested in right now. The beans and special for me, please.'

'Me too,' echoed Charlie, patting his stomach. 'If in doubt, go for the special every time.'

'You know,' Ellie mused, 'I'm almost tempted to order the special myself. Hmmm – *Rates Sa Pobla* – there's a sort of exotic ring about it. Can't you ask what's in it?'

'I did, and all I can tell you is that it's a recipe from the town of Sa Pobla in the north of the island.' I shook my head and added, 'No, I'd have the pork, if I were you. All those oak trees around these parts mean masses of acorns for the pigs to eat. The pork's bound to be fabulous. Always is in Mallorca. In fact, I've eaten so much pork recently that I'm surprised I haven't turned into a pig.'

'Don't push me for an opinion on that one,' Ellie muttered. 'Just order me the pork.'

As is normal with Mallorcan *Menús del Día*, the first courses were unashamedly filling and wholesome, with no pretence at being merely dainty little appetisers. They were built to satisfy the working man with an empty belly, not to smudge the middle of a plate like some *nouvelle cuisine* designer scraps.

Our curiously-named 'American Rice' had been moulded into plate-size volcanoes concealed under herby, red lava flows of a rich sauce, and the boys' food would have made even the great H.J. Heinz look to his laurels. Our sons guzzled their way through the overflowing platefuls with gusto, and the waitress routinely ladled out second helpings without having to be asked.

From what we could detect during the short time that the food was on their plates, the boys' second course, the *Rates Sa Pobla*, was a casserole of little chunks of browned white meat, done in a substantial garlicky gravy of chopped leeks, red peppers and baby tomatoes. There was no need to ask if it was good.

Our grilled pork cutlets were so tender that they cut as easily as sponge cake – the ultimate testimonial to the gastronomic significance of the humble acorn, when processed by a free-range Mallorcan *porc*.

While politely declining the *postre* of oranges, I thanked the waitress for her discerning service – at which she smiled down on the silent heads of our sated sons – and I asked her to convey our congratulations to the kitchen on their wonderful pork-based dishes.

The end of the year, she explained, was the season of the *matances*, the killing of the pig on the little farms, so there was always plenty of excellent fresh pork and associated *productos* available at this time. She returned to the kitchen and reappeared a moment later with a small plate of *Taronja Confitada* – seductive strips of candied orange peel – with the compliments of the chef, who hoped that these sweet and tangy titbits would appeal to *los señores* more than the ordinary whole oranges which had been on his *menú modesto* today. The only *modesto* constituent of this meal was the price.

'I take it that you enjoyed your main course, lads,' I remarked once we were on the road home. 'You certainly polished it off quickly enough.'

'Ahhh, it was tremendous,' Sandy enthused dreamily, sprawling in a stuffed heap in the back of the car.

'Dead right,' grunted Charlie, undoing the top button of his jeans. 'Different class. I've never tasted chicken like that before. Mmm-wah!'

'Only one thing, lads,' I said. 'That wasn't chicken in your main course.'

'Well, rabbit then,' mumbled Sandy disinterestedly. 'I thought it might have been rabbit, that special flavour.'

'Yeah, whatever,' agreed Charlie. 'Chicken, rabbit – no pressure. Different class.'

'No, not rabbit either,' I advised calmly. 'You see, I checked with the waitress when I was paying the bill, and she confirmed that *Greixera de Rates Sa Pobla* means literally "Casserole of Sa Pobla Rats".'

– NINE –

OLÉ HOGMANAY

As the last day of the year came round, I realised that we hadn't seen Jock Burns since that chance meeting at the airport on Christmas Eve, and as we had promised to get together during the festive season, I ventured to ask Señor Bonet if it might be possible for our countryman and his wife to join us at the grand occasion of his New Year's Eve Party at the Heladería Ca'n Toneta. They would not impose upon his hospitality, I stressed, and would only be present for the latter stages of the festivities – after Jock had finished his hotel gig along the coast, in fact.

My hesitant request was met with a typically magnanimous: '*Mi casa es su casa, amigo*, and your friends are mine!'

Jock and Meg were duly invited, and we looked forward to having some long-overdue fun in their company.

* * * * * * * * *

Señor Bonet's ice cream parlour was looking surprisingly festive and even attractive, in a disturbing sort of way – a bit like an ageing, has-been actress who had been crudely tarted up for a rare night out on the town. Boas of tinsel and coloured crêpe paper had been draped over the archways, and the bank of ancient spotlights glared down, after years of hibernation, through the theatrical smell of burning dust onto the little bandstand. The bandstand's curtains were tied back in lop-sided folds to reveal the one-time ballroom's gaudy bosom-brooch of a lame-jacketed 'orchestra', tonight reduced through financial expedience to a keyboard player and a drummer, slumped silently in front of a huge cracked mirror which, judging by the intricate mouldings of its mouldy gilt frame, had probably spent the prime of its long lifetime as an overmantle in the salon of some grandiose Palma mansion. What glorious scenes of finery and grace it must once have reflected in the extravagant brilliance of infinite rows of crystal chandeliers. But not tonight.

Bulky, low-consumption light bulbs had been squeezed into the frosted glass shades of the ceiling pendants, the ribbed, cylindrical lamps bulging like bloated tongues from the gaping mouths of strangled electric Hydras and filling the room with all the bald, unflattering dazzle of cinema cleaning lights. The lines of white plastic tables and handsome tapestry chairs had been rearranged round the perimeter of the hall, the cane screens swept aside to leave an area of dance floor with an uninterrupted view of the spaghetti-wired backs of the various gaming machines and refrigerator display cabinets, lined up like a static company of robot sentries guarding the approaches to the long bar. His spreading form forced uncomfortably into a skimpy tuxedo which clearly dated from the glory years of his ballroom, Señor Bonet stood behind the bar, proudly welcoming everyone with shouts of '*Benvinguts*' and, as it had become apparent that the 'orchestra' was taking a break, he purposefully adjusted the volume of the cassette player on the gantry, giving Julio Iglesias an opportunity to perform 'To All The Girls I've Loved Before' from that divine position beneath the statuette of the Blessed Virgin.

Señora Bonet, a seldom-seen but pleasantly-disposed lady of delicate build and with a permanent expression of resigned boredom on her kitchen-blanched face, had been released from her back room duties for the evening and, demurely attired in her little-used front shop black dress, was stationed in a corner behind a row of trestle tables that were flexing under a fabulous array of freshly-prepared eats. Banks of hotplates were submerged under a steaming lily pond of round earthenware *greixoneras* containing all kinds of *tapas* – traditional savoury titbits, once served in Spanish hostelries as something to pick at while drinking, but now more often eaten in meal-sized quantities in special *tapa* bars where drinking has been relegated to second place. Well, almost.

Señora Bonet's *tapas* selection was truly magnificent – a bewitching arrangement of colours, aromas and flavours which constituted a celebration of the bounteous gifts of the Mallorcan land and the Mediterranean sea: gleaming red prawns and *langostines* grilled in olive oil and lemon juice; golden deep-fried squid rings; potato chunks smothered in garlic mayonnaise; little balls of minced ham and veal bubbling in parsley sauce; silvery baby eels simmered in white wine; tender lamb's kidneys gently stewed in sherry; black *butifarrón* and white *butifarrón* sausages oozing the exotic savour of cinnamon and cumin; even plump dates rolled in best *serrano* ham.

As an alternative, or perhaps even as a follower to these 'nibbles', Señora Bonet had made several huge *greixoneras* of *Sopes de Matances* – not really a soup at all, but a wholesome stew as thick as a mattress and a favourite Mallorcan dish at this season of the *matances*, the celebrated killing of the family pig.

'If there isn't enough space on the floor, they'll be able to dance on that stuff,' Ellie quipped.

The star of Señora Bonet's culinary show was undoubtedly the *Porcella Asada*, a princely suckling pig 'sacrificed' at three weeks of age and solemnly basted with olive oil and lemon juice while being roasted to crackly, golden-brown perfection to finally lie in scrumptious state on a bed of roast potatoes and green beans.

By the piglet's side, lean shoulders and legs of baby Mallorcan lamb sizzled in attendance, flanked by an army of little birds baked in snappy, skewered uniforms of crunchy bacon.

'Quails?' asked Ellie, peering suspiciously at the platoons of tiny, winged soldiers.

'Yes,' I lied, knowing perfectly well that they were more likely to be thrushes. Mind you, I thought to myself, what difference did it make anyway, except that – depending on your ear – one variety of bird had a nicer voice than the other? The birds certainly couldn't have given a hoot now. Quail, thrush, crow or dodo – it doesn't matter what name you've been given when you're lying on your back with your feet in the air waiting to be eaten.

Finally, *greixoneras* of *Escaldums* – turkey pieces bubbling in a creamy sauce – led to bowls of dressed salad and heaped platters of saffron rice which, in turn, gave way to a rearguard made up of trays lined with countless ranks of *Bunyols* (little Mallorcan doughnuts), hissing hot from the fryer and liberally dusted with icing sugar.

The tables round the hall had been decked with baskets of country bread, dishes of pickled olives, jugs of wine, bottles of *gasiosa* lemonade and jars of iced water to keep everyone sustained until Señor Bonet had decided that the invited company was complete and it was time to eat. Then, after catching our attention by rapidly switching off and on the ceiling lights (with the resultant popped expiration of half a dozen low-consumption lamps), he genially beckoned his guests to converge upon the row of trestle tables where he and the two kitchen maids, both well dolled up in scanty little sequinned numbers for the *fiesta*, had joined Señora Bonet to help dispense the treats from her *banquete grande*.

When the ensuing free-for-all had finally sorted itself out and we had all successfully yelled our preferences to the harassed buffet tenders, brimming *platos* were conveyed back to the tables where the feast got seriously under way, the revellers eating, shouting and laughing in victorious competition with the taped tones of Plácido Domingo, who had now replaced Julio Iglesias beneath the Holy Virgin of The Gantry.

The two members of the 'orchestra', meanwhile, had taken up their natural slouched positions at the now-deserted bar, from where – safely ensconced behind their self-made security screen of beer fumes and nostril-exhaled smoke –

they could cynically survey through their dark glasses the alien food fixation of the assembled Earth People.

As the meal wore on and the wine flowed more freely, the volume of table talk increased as usual, providing a welcome cover for our total lack of comprehension of the bedlam of conversation in *mallorquín* which surrounded us. A smile, a nod and a mimed 'sí' seemed to suffice when a question or observation was directed our way.

Now that the rush at the buffet was over, the two kitchen maids had been seconded to the drinks department and were wobbling on outrageously high heels between the bar and the tables, dodging bearers of second helpings, replacing empty wine jugs with full ones, and revelling in the furtive leering of the hen-pecked husbands who strained to catch a surreptitious eyeful of that extra inch of cleavage and thigh exposed by the girls with controlled precision while stretching for empties over the crowded tables. The fact that the kitchen maids had legs only marginally more shapely than the skewered thrushes' – and were no better endowed in the cleavage department either – did not matter to the leerers. It was the tantalisingly limited exposure of flesh against sequins that got them going, and their fascination grew with the intake of free wine, the boldness of their ogling increasing apace, until one particularly miffed wife ended all the fun by simultaneously tripping up the kitchen maids and spilling a jug of iced water into her husband's over-heated lap.

The din of the resultant fracas and the sight of the kitchen maids sliding over the floor on their faces seemed to serve as a signal to the two musicians that the party had now warmed up sufficiently to warrant some live music. They proceeded to the bandstand.

The keyboard player looked at least one generation older than the drummer, and this was reflected in the duo's treatment of the first medley – a selection of Glen Miller hits from the Forties, the tunes executed on the keyboard in a style that owed more to the *paso doble* than swing, while the gum-chewing drummer thrashed out a notably un-hip beat that could best be described as a poor Ringo Starr impersonation of Gene Krupa on an off night. Nevertheless, several determined couples took to the floor, the music inspiring routines ranging from flashy, quick-stepped fish tails through vintage jitterbugging to the hand jive and some 'bad' get-on-down boogie. We even noticed old Maria Bauzá doing a rheumatic solo version of an ancient Mallorcan folk dance to the melody of 'Chattanooga Choo-choo' in the no-go area outside the gents' toilet. When it came to cutting a rug, she evidently wasn't just a mere conversation behind, but an entire epoch.

No matter; the musicians had done the business. They had got the audience in the mood, and the party was up and running.

The only immediate neighbours we hadn't seen were the Ferrers, it probably being a mite infra dig – in their own opinion – to be seen at a party with the common hoi polloi of Andratx. It would be more in keeping with their developed image, no doubt, to be swanning around at some swank do for civic dignitaries in Palma. And the best of luck to them. Most of our favourite local characters were right there in the Heladería, and it was great to see them letting their hair down.

Old Jaume had claimed the centre of the dance floor now, and was delighting his two giggling granddaughters with his own peculiar version of some trendy dance or other that he had doubtless observed during his career as a waiter at the Hotel Son Vida – his full belly bouncing up and down above gamely-prancing knees, his elbows waving wildly, his horn-rimmed specs beating perfect time on the end of his nose, and an ear-to-ear grin on his crimson, perspiring face.

'That's the 'Mashed Potato' he's doing there,' Ellie yelled in my ear.

'No, just look at the arm movements,' I argued. 'That's the 'Funky Chicken', if I ever saw it.'

We watched Jaume's gyrating globular body for a few moments more then agreed in nodding unison:

'Yeah, it's the 'Funky Potato'.'

'Anyhow,' Ellie laughed, 'it's nice to see Jaume enjoying himself with his family. He's obviously thrilled to have them back home for the holidays. Mm-mm, that's really nice.'

'Certainly is,' I agreed, raising my voice against the amplified roars of 'PENNSYLVANIA SIX – FIVE – O-O-O!' thundering from the bandstand. 'And talking about families, where have our two heroes disappeared to?'

Ellie pointed first towards the entrance to the back shop, where Charlie and young Toni had teamed up and were busy sniggering away at something going on through the doorway. Then she nodded towards the bar, where Sandy was engaged in deep conversation with the Andratx football team's version of Maradona, whom we hadn't seen since our unforgettable dinner at Pere Pau's tiny eating place round the corner on our first night in Andratx.

My eyes wandered along to the other end of the bar, and there, leaning in characteristic cross-legged pose against the switched-off pinball machine by the door, was old Pep, looking fairly switched-on himself in his stepping-out gear of best black beret, new polka-dot neckerchief and specially polished leather bomber jacket, his head inclined backwards as he viewed the merry-making with contrived disdain through eyes half-closed against his stinging exhalations of venomous *cigarrillo* smoke. I waved to him across the hall, and he winked an eye almost imperceptibly in response. This public Pep was Mr Cool personified.

I felt a nudge in my back, and someone shouted in my ear:

'Old bloody damn bastard!'

I spun round to see the little man who had given me the definitive explanation for the car-up-the-almond-tree mystery when I was buying a newspaper in Puerto Andratx a few weeks earlier.

He gestured in old Pep's direction and repeated in his best English bad language, 'Old bloody damn bastard! Still be thinking he running the bloody bugger town. Is ridickliss.' He chortled mischievously and pulled up a chair beside me, bowing stiffly to Ellie and offering her his hand. 'How you doing, lady? I no been seeing you there, so please be excusing my language. One Pakistani bloke been learning it to me when I been working in one car factory in Coventry many years. I am Jordi. Jordi Beltran Nicolau, carpenter the boats and many expert.'

We shook hands and introduced ourselves while Jordi smiled uninhibitedly and uttered little chuckles after repeating each of our names:

'Jordi . . . Peter . . . Jordi . . . Ellie. Bloody 'ell.'

He was having a great time, his alert Mallorcan eyes shining from thin, well lived-in features beneath a thick shock of greying hair.

'So you lived in England for a while, Jordi?' I enquired.

'Damn sixteen years, I tell you,' he declared, proudly sticking out his skinny chest. He crossed his legs, the outline of his bony knees showing through his well lived-in trousers. 'My wife, she being English. Oh yes,' he added, flicking a trace of cigarette ash from the sleeve of his faded cotton jacket. 'And my kids too. Three very good ones. All girl ones, and all been being born in England, oh yes.'

'And you all live in Mallorca now?' asked Ellie.

Jordi continued to smile, but a trace of sadness had dulled his eyes. 'No, only me now, lady. The wife and kids been coming back here with me for living five years ago. I having one old house with one big palm tree, very beauty in Andratx, oh yes. But the damn summer weather here been being too bloody hot for the wife.' He forced out a plucky laugh. 'I tell you, one year in Andratx and she been going back in England with the kids. All in bloody damn Coventry again, looking after the old grandfather – like him, Pep – old bloody damn bastard.'

I had little doubt that further stories of Jordi's personal life would have made interesting listening, and although he wouldn't have needed much persuasion to tell all, I didn't want to appear too nosey, so I gladly grabbed the opportunity to change the subject.

'Ah yes – old Pep,' I said. 'Quite a character, isn't he? He's our neighbour, you know. We live up at –'

'Oh yes, you been buying the *finca* of Ferrer – bloody damn mayor of Palma, he thinking himself. Is ridickliss. Oh yes, I can knowing all that. I tell you, I can knowing everybodies and all happening in Andratx. With the speaking good English, I can tell you everything, bloody 'ell!' Jordi re-crossed his legs and took a sip of a strange, watery-brown liquid from his glass – a lethal-looking concoction which he proudly explained was a mixture of Minorcan gin and 'Palo', a local liqueur made from quinine . . . and the roots of the Maria weed. 'Cheers!' he beamed. 'My very good health down the hatches!'

'Yes – your, ehm, very good health indeed, Jordi,' I reciprocated. 'All the best.'

I sipped my wine while Jordi lit a cigarette.

'Spanish ones,' he grinned, flashing the packet at me. 'English fags been giving me pipes like the razor's blade, and the mouth with cow's arse flavourings. Is bloody ridickliss.'

Ellie excused herself politely and escaped to the ladies' room.

'That thing you mentioned about old Pep thinking he still runs the town, Jordi,' I prompted, my curiosity getting the better of me. 'Eh . . . what exactly did you mean?'

'Oh, he been being big sheet in Andratx many years ago, oh yes.' Jordi leaned in close to make sure I didn't miss any of the hot details. 'Pep been being big sheet on the council here, and your other neighbour, Ferrer, been being his assisting. Yes, bloody 'ell, Ferrer been being the assisting of Pep! And I tell you, Pep been being the boyfriend of the Señora Francisca since she been being a girl – very beauty in them days and many pokings with Pep in the fruit trees one summer.'

Jordi slapped his knee and guffawed, clearly delighting in the memory of this little gem of past scandal. He went on to tell me that, as Francisca had only been about thirteen and Pep more than twice her age when old Paco discovered the intimacy of his only daughter's relationship with this rascally man of the world, there had been such a furore locally that Pep had been forced to take up a government post in Cuba until the dust settled. When Pep returned to Andratx a few years later, not only had Tomàs Ferrer been promoted into Pep's top job with the council, but he had also married the lovely Francisca. Pep was devastated. He married another girl on the rebound, but it didn't work out. He started to hit the bottle heavily, often being seen lying about the streets at night after drinking himself stupid in the town's bars.

From being one of the most respected men in the area, Pep had become the town drunk, and then had to suffer the added pain of seeing his former assistant climbing the ladder of success with the cherished Francisca by his side. With his life in ruins and all his money gone, Pep had returned to the little *finca* in the valley, deserted and overgrown since the death of his parents while he was in Cuba. He then lived the life of a recluse for years, never touching alcohol again, but working day and night with the oldest of derelict equipment to restore the farm to productivity and to build his herd of sheep – and all this while being subjected to the agony of seeing the ever more affluent Tomàs and Francisca returning to visit her parent's *finca* just over the lane every weekend.

But time healed the wounds, and little by little, Pep began to join in the daily life of the little community again, although those lost years had changed him forever. He was no longer the dashing administrator, the happy-go-lucky local somebody with the world on a string and a young girl's heart in his pocket, but a tough old nut in an ill-tempered shell, with an irritating air of superiority that some said was just a kind of madness caused by his legendary drinking habits of years gone by, while others maintained that he was nothing but a grumpy old miser, sitting alone, brooding over a fortune that he had salted away peseta by peseta during half a lifetime of living no better than his mule.

All Jordi knew for certain was that Pep reminded him of his wife's father back in Coventry, who was certainly an old, bloody, mean, damn bastard, interested in nobody but himself.

'I tell you,' he said, standing up to go and further exercise his conversational energies on someone he had just noticed over at the bar, 'the two of them's be being like two piss in a pods. Is bloody ridickliss.'

Whether or not Jordi's story about Pep was entirely true, I didn't know, but it would certainly have explained many of the puzzling aspects of our enigmatic old neighbour's character traits, as well as his low regard for Tomàs Ferrer. Perhaps some of Maria Bauzá's bitterness towards Francisca Ferrer also stemmed from those alleged promiscuous activities in the orchard many decades ago. Had Maria been the one to witness the forbidden fruit being eaten by Francisca and Pep in the neighbouring orange grove? Had Maria spilled the beans about the fruity goings-on to old Paco in a fit of jealousy because she maybe fancied young Pep herself? Perhaps one of Pep's official duties had been the administration of shared water rights, so had he used his position as a front for giving the pubescent Francisca some unofficial demonstrations in the art of productive hose-pipe handling behind the well? Yes, of course – that would also account for old Maria's fervid aversion to shared wells!

A finger jabbing impatiently into my back brought my wild fantasising to an abrupt end. It was old Rafael – face all polished to a high gloss, breath reeking of wine, and clothes still humming of goats. There was also a new, tart smell about him, which he cheerfully volunteered came from the goats' milk yoghurt that he had used to plaster down his hair. Drove the *muchachas* crazy, he told me. He shook my hand cordially, them ambled off with a self-assured swagger to subject some unfortunate old dear to insanity by asphyxiation on the dance floor.

The kitchen maids' shuttle service of free wine had now ceased, so I suggested to Ellie, who had just returned to her seat, that we make our way over to the bar, as the form now appeared to be: If you want any more to drink, buy it. Señor Bonet had been a generous host, but business was business after all.

Sandy was still huddled in conversation with the Andratx Maradona, and we eavesdropped on them while we waited to be served at the crowded bar.

'But I never played football at international level for the Scottish Schoolboys,' pleaded Sandy. 'Honest – I was lucky to make it into the school second team . . . and that was usually as a sub.'

'No worries, mate. The La Real team are well desperate for a new sweeper. No shit, please believe, OK? So I gonna tell their manager – he's a bleedin' ned, OK? – that I come across this brilliant kid – that's you, fuckin' genius – and you been playin' for the Scottish Schoolboys plenty at Wembley, OK?'

'Hampden. Scotland plays at Hampden, not Wembley.'

'Screw Hampden, mate. Nobody never heard of bleedin' Hampden. Everybody heard of Wembley, OK? So I give all this Wembley shit to the La Real twonk and you're in. Got me? Yeah, no danger, mate. Nice one.'

The streetwise Maradona must then have sensed that his delicate contractual negotiations were being listened in to. He turned round with a black look on his face, which was instantly transformed into his mirror-rehearsed, one-sided smile when he recognised me.

'Hey, daddy, how they hangin', mate?' he yelled, punching my shoulder. 'Nice to see ya. I just been fixing your kid to sign for a team up the island there – La Real – loada crap, but you gotta start somewhere, OK? If he gonna make the grade with La Real this season, I gonna fix him a contract for Andratx next season. No danger, alright? Hey, missus, you lookin' t'riffic,' he beamed, noticing Ellie by my side and grasping her in an over-familiar hug. 'Wow, doll face – what I give for a mamma like you!'

'Oooh! Oh! It's nice to see you again too, Mister ehm . . .' squirmed Ellie. 'And how is your young lady? Not with you tonight?'

Maradona raised his eyebrows and, without moving his head, turned his eyes disinterestedly towards the bar, where his doting groupie was standing gazing moodily at him, her hands clasped over the scarcely perceivable beginnings of a distended tummy.

'Yeah, she always hangin' around,' he drawled. 'Right up the bleedin' spout too. Well sprogged, mate.'

'Oh, that's nice,' said Ellie with an uneasy smile. 'You're going to be a father then?'

'Yeah, me and the rest, missus,' he smirked, taking a swig from his bottle of beer. 'If she wanna name that chavvi after its father, she gonna have to call it Andratx United, OK? No shit, she had her hands on more naked bodies in the team this season than the bleedin' club physio. Yeah, no fuckin' danger, darlin'.'

Maradona's spoken English, not to mention his social graces, were a continuous credit to his tutoring by the Brit lager louts who frequented the beach bar where he worked, and although he enjoyed giving the impression that he was also of that profligate ilk, we fancied that sterner influences would ultimately prevail, and it probably wouldn't be too long before he, too, was doing the 'right thing' by starring in his own shotgun wedding reception at the Restaurante Son Berga.

The lights flashed on and off again, a drum roll like a tin trash can falling down a flight of stone stairs rattled out from the bandstand, and Señor Bonet strode onto the stage with the slick gait of the truly classy ballroom manager, his stride only slightly restricted by the gusset of his undersized trousers biting into his crotch. He whipped the microphone off its stand, tapped it and blew into it a few times like a real professional, then, amid ear-ripping howls and whistles of feedback from the loudspeakers, he introduced the cabaret. With a Thespian gesture towards the door to the back shop, he announced:

'*Señoras y caballeros . . . El Flamenco Famoso de LAS HERMANAS DE GRANADA-A-A!*'

The 'orchestra' struck up the introduction to 'Sol y Sombra', that majestic overture which reverberates round a thousand bull-rings every weekend, and after an agonisingly long pause, during which some of the die-hards at the bar started to order drinks again, the Sisters of Granada Flamenco troupe – all two of them – swept past us in their figure-hugging, flared frocks and launched themselves into a full-blooded, high-heeled assault on the Heladería's floor tiles.

'So that's what all the sniggering from Charlie and Toni was all about,' I murmured through a knowing smile.

Ellie drew her brow into a puzzled frown. 'The Flamenco dancers?'

'That's right. Las Hermanas de Granada are none other than the kitchen maids, and those two young pervs have been standing at the door to the back shop watching them change their clothes.'

'The dirty little dogs! Charlie's going to get it for this.'

'I shouldn't bother, dear. Boys will be boys, and anyway, I don't think there's much about that pair of fandangoing match-stick ladies to get the lads hooked on the Peeping Tom habit.'

'Don't you believe it. Look at old Rafael, for instance. His eyes are just about popping out of his head every time the girls do a twirl and show a bit of leg.'

'Hmmm. Must have been on the wormy orange juice again, eh? You know, I think we should start bottling that stuff. If Rafael is anything to go by, we could make our fortune selling it as a passion potion.'

'You said it. Just look at him now, for heaven's sake!'

Rafael had flipped. The sound of his native Andalusian music and the blood-stirring spectacle of the gypsy dancing had been too much. It had all gone to his head – with a little aid from a gutful of free wine – and he had joined the kitchen maids on the floor, podgy fingers snapping above his head, rubber-soled sneakers stomping on the hard tiles with all the percussive precision of two wet fish landing on a marble slab. He bent his knees and pushed his pelvis forward seductively as he rotated ever so slowly, his chin tucked in and his eyes lowered threateningly, displaying all the controlled aggression and bridled menace of a little fat matador facing a killer hedgehog.

The place was in an uproar. Shouts of *'OLÉ!'* rang out and the music grew louder and faster, beads of sweat appearing on Rafael's brow as his clapped-out body protested against the impossibility of this grotesque ballet which it was being forced to perform. Rafael's feet may only have been shuffling on the floor of a stuffy ice cream parlour in Andratx, but his befuddled brain cells were in orbit somewhere above Granada, believing that he was dancing among the stars like a spinning Adonis amidst a host of sultry Andalusian angels in the hot, perfumed air of the Sierra Nevada.

The beads of sweat turned into streams and the streams into rivers, washing Rafael's lactic hair lotion down over his face and onto his sodden shirt. The kitchen maids bolted the course in tears, probably piqued by this little *borracho* blowing their one big break in showbusiness, but more likely gassed by the stench of putrid yoghurt mixed with steaming goat fumes.

Rafael was left on his own on the dance floor, a lone Romany prince strutting his stuff round the camp fire while the gypsy guitars strummed feverishly and his adoring *compañeros* shouted and clapped their encouragement – inspiring even more magnificent turns of Flamenco improvisation from his lithe, rhythm-charged limbs. His eyes were closed in ecstasy, his mouth open now, wailing the plaintive, trembling chants of his Moorish forefathers, his head thrown back and shaking in uncontrolled frenzy. Rafael was sent.

Back in the real world, the 'orchestra' had already abandoned the stage for another interlude at the bar, and a wine-weary old goatherd was staggering about the floor, moaning incoherently in a solo stupor, with the rest of the guests doubled up in fits of laughter watching his baggy trousers slipping further and further down with each unsteady rotation of his sagging frame. Just as the trousers settled round his ankles, revealing a pair of wrinkled, beige combinations complete with buttoned back flap, a well-timed signal from Señor Bonet brought two burly young men out of the crowd, grabbing Rafael before he fell over, and carting him off to the seclusion of the back patio, where the cool night air would either bring him back to his senses or bring on a fatal bout of pneumonia.

Señor Bonet wasn't fussy either way. He was already back on stage introducing the next act, a group of young folk dancers dressed in traditional Mallorcan costumes; the girls in face-hugging white head-dresses which flared out into lace-trimmed borders at the shoulder, black, long-sleeved, fitted bodices, and long, full skirts of various colours, embellished with dainty white aprons; the boys in full-sleeved white shirts, little black waistcoats, and billowing *pantalones* of colourful striped cotton gathered in below the knee over long white socks. Their music was provided by two musicians wearing Mallorcan peasant dress – one beating a *tambor*, a simple drum hanging from the shoulder on a leather sash, the other playing the *xiramías*, little Mallorcan bagpipes which sing with a quieter, silkier voice than their spine-chilling Scottish cousins.

Gentle, lilting melodies – echoes of Arabia distilled through a millennium of island folklore – soon filled the Heladería, the dancers stepping and turning, skipping and whirling in a graceful *parado*, delighting their audience with the joyful exuberance of youth. It was truly a sight to warm the Mallorcan heart, and no one could have been more thrilled than old Maria Bauzá, standing with tears of happiness in her eyes, her five teeth glinting in a spot-lit smile as she swayed her stooped shoulders and swung the raised hands of her two great-granddaughters from side to side in time to the music. The strains of those ancient airs and the spectacle of the young people rejoicing in the familiar patterns of the dance had transported her back to her beloved old days. She was a girl herself once more, dancing in the summer fields, her bare feet carrying her supple young body over the warm ground as lightly as a leaf caught in a gentle breeze, the strong hands of her *novio* holding her tiny waist as they swirled and circled, lost in the carefree dreams of adolescence.

Just then, our attention was drawn to the main door by a strident vocal fanfare:

'YOO-EE-EE! OH, YOO-EE-HEE-EE!'

It was Jock's wife, Meg, making her grand entrance into the Heladería, her buxom figure clothed in a flowing robe of so many colours that it might have been made from the combined flags of the United Nations. There was a grin on her face as wide as a slice of melon, a tousled garland of streamers dangling from her blonde hair, a brilliant sparkle in her eyes generated by the unquenchable jollity of the seasoned raver, two silver balloons in one hand, and a half-empty bottle of champagne in the other. Rent-a-party had arrived.

When it came to having a good time, Meg was in the major league, and if years of eating, drinking and being merry had added a few inches to her waistline, so what? She had a built-in vivacity that was indestructible and her exuberant approach to life was reflected in her face, which was as radiant and stunningly attractive as it had ever been in her pre-Pimms girlhood. Meg was a showstopper, and all eyes were on her as she snaked towards us like a one-woman Rio Carnival.

'HI-EE-EE! WELL, HELLO THERE, FLOWERS!' she called out, smothering Ellie and me in cuddles and kisses, then giving the proceedings a quick once-over. 'My, my – this is a cosy little party. When does the fun start, petal?'

'It won't be long, not now that you're here,' I kidded.

Meg let out an unrestrained cackle of laughter that drowned out the music of the *xiramías* for a few moments, causing looks of near panic among the folk dancers.

'And how about the old-time dancing?' she remarked aloud. 'Oh dear, we'll have to do something about this. I mean, be fair – that stuff went out with Noah. We're gonna have to put some life into this lot.'

'Ahem, how come you're here on your own, Meg?' asked Ellie, patently worried that Meg was about to throw herself twisting and shouting into the well-ordered formation of dancers. 'Where's Jock?'

'I haven't managed to ditch him, if that's what you mean. No such luck. Nah, he's only parking the car, but when he got an earful of this minstrel music coming out of here, he said it gave him a great idea, so who knows what he's up to? Whatever it is, it'll be an embarrassment – you can bet on that.'

I heard Pep rasping out a faked cough just behind me.

'Oh, I'm sorry, Meg. This is Pep, our neighbour from up the valley,' I said, quite surprised that the old fellow had crept up on us so quietly. Contrary to what I would have expected, Meg's over-the-top presence had clearly aroused his interest, and had begun to rekindle the merest glimmer of the romantic flame extinguished by the unfaithful Francisca on his return from Cuba almost forty years ago.

Pep straightened his neckerchief and doffed his beret. He took Meg's hand and raised it to his lips, even remembering – at the very last second – to remove the ever-present cigarette from the corner of his mouth. *'Encantado, madame,'* he whispered throatily. Pep was smitten.

'This old hell-raiser looks as if he hasn't had a good laugh since he escaped from the Inquisition,' joked Meg, flashing Pep one of her most alluring smiles. 'But never mind. I'll soon get his motor fired up.' She pinched Pep's cheek and wrinkled her nose at him. 'Won't I, flower?'

A grunt and a scowl was Pep's only response.

This was shaping up to be a classic example of the irresistible force versus the immovable object. It promised to be fun!

The kitchen maids, sufficiently recovered from their humiliating and short career as Las Hermanas de Granada, had reappeared in their sequinned strumpet dresses to distribute bunches of grapes throughout the hall, and these were now being divided up by the guests as Señor Bonet called a halt to the folk dancing and switched on a TV set behind the bar.

'Here, you'll need these,' Meg shouted, lobbing us a handful of grapes apiece. 'It's the custom here. You have to swallow a grape on each on the twelve strokes of midnight. It's supposed to bring you luck in the New Year. *If* you can do it.'

The TV flickered to life, showing scenes of merrymaking crowds eagerly awaiting the advent of New Year outside the old City Hall in Palma. Señor Bonet turned up the volume right on cue, and as the bells rang out, the old arches of the Heladería resounded with the sound of chomping and choking from the grape-chewing followers of tradition, and with the cheering, laughter and cork-popping of the more fundamental fun-seekers who regarded the quaffing of champagne direct from communal bottles as traditional enough for them.

We made a brave attempt at upholding the old custom, but after gamely stuffing away five or six grapes, we were obliged to desert to the camp of the new traditionalists. We filled our glasses from Meg's bottle, toasted the infant year, and lost ourselves in that over-emotional mêlée of handshaking, kissing, hugging and back-slapping engendered by the benevolence towards the whole of mankind that mysteriously eclipses our entrenched prejudices for those few maudlin minutes at the birth of each New Year.

The 'orchestra' played 'Auld Lang Syne' and the bonhomie became almost unbearable. Old ladies sniffed tears of nostalgia into their hankies, old men stood nose-to-nose in bleary brotherly love, other people joined hands round the dance floor in the universally accepted way and sang strange phonetic versions of Rabbie Burns' hallowed lyrics, while some opportunist teenage couples made the most of this emotionally demonstrative interlude by having a good old grope at one another in some of the darker corners of the room.

As the shouts of *'Molts d'anys'* and *'Feliz año nuevo'* rang out at the end of 'Auld Lang Syne', a familiar yet out-of-place sound cut through the barrage of bursting balloons and the squawking of paper hooters. It was the ghostly sound of distant bagpipes – not the little humming-bird drone of the *xiramías*, but the full, blood-curdling skirl of the Scottish war pipes. The street door flew open, and in marched the phantom piper: Jock, deliberately over-waggling his hips to exaggerate the pleated swinging of an ill-fitting kilt which drooped so low at the back that it almost touched the tops of his short socks. His left elbow pumped the tartan bag like fury, his chest heaving fit to burst, his ruddy cheeks inflating and deflating like a pair of amorous bullfrogs as he squeezed and coaxed the squealing, wailing notes out of the chanter.

'See what I mean?' yelled Meg, putting her hands over her eyes. 'Jock and embarrassment are identical twins.'

'But where did he get the bagpipes and kilt gear?' I shouted.

'From a Scottish holidaymaker he met in the hotel tonight. He was skint, so Jock offered him fifty quid for the lot, and the bloke accepted, more's the pity.'

'But I never knew that Jock could play the pipes.'

'And by the sound of things, I don't think he can, although he says he played them in the Boys' Brigade when he was a kid.' Meg shrugged her shoulders. 'Be your own judge.'

Although Jock certainly would not have won any prizes for his bagpiping technique (which sounded, at best, a little rusty), this didn't prevent the keyboard player and his drummer from latching onto 'where Jock was at', and before Señor Bonet's customers knew what had hit them, they were being treated to a wild, spontaneous rave-up from the most unlikely of musical pick-up groups.

Jock was like a man possessed, blowing up a storm of reels and jigs, with the *xiramía* man doing his best to play along – albeit in another key – and the other *músicos* tearing into this freaky jam session with all the unrestrained gusto of turned-on musicians everywhere when they plug into that mystical source of harmonious inspiration and unbridled jollity known only to their closed brotherhood . . . and maybe to a few tuneful distillery workers. They were having a ball, and the crowd loved it.

'WOW-EE-EE!' whooped Meg. 'It's party time! Come on then, flower. Let's burn it up!' She grabbed Pep by both hands and dragged him onto the floor. Without giving the bewildered old codger a chance to object, Meg crooked her arm in his and whirled him round and round in a dizzy spin. It was all Pep could do to keep his beret on.

'I think that pair are going to need some back-up when they let go of each other,' I shouted in Ellie's ear.

'Well, I'm game, if you are. Let's go!'

We took to the floor and hurled ourselves into an impromptu Foursome Reel with Meg and Pep, zigzagging up and down the hall, and twirling and birling each other until our arms ached. Jock and his motley band redoubled their wild outpouring of improvised ceilidh music, the pipers blasting out a gale of ad-lib Celtic cadenzas, the keyboard player vamping away like a turbo-charged steam hammer, and the two drummers belting out the beat as if their lives depended on it. The fever of the red-hot rhythm spread like wildfire, and soon the little dance floor was a swarm of bodies charging around in a madcap Highland Fling – arms arched above their heads, pointed toes pas-de-basquing, and flying feet high-kicking.

Ever the considerate ballroom proprietor, Señor Bonet switched the ceiling fans to full speed for the comfort of the dancers, urgently sending his wife on stage with a long pole to push-start a reluctant propeller above the perspiring musicians.

Jock had his back to Señora Bonet as she jabbed away at the stubborn fan, and he was so engrossed in his playing that he was completely unaware of the bottom end of her unwieldy pole inadvertently catching in the drooping hem of his kilt and lifting it right up to hook on the buckle of his sporran strap at the small of his back. Being a true Scotsman, Jock had left his underpants in the car, so when he turned round to signal the approaching end of the first set of tunes to his sidemen, the audience was treated to a full view of the twin moons of his naked behind.

An almighty roar of laughter went up as everyone stopped dancing to stand pointing and shouting catcalls of derision at Jock's unwitting act of indecent exposure. Mistaking this mass mockery for a spontaneous show of adulation, Jock decided to keep playing when the other musicians stopped. The show-off in him had taken over completely, and he launched into a bizarre bagpipe version of the Samba, turning his back on the crowd and wiggling what he innocently believed to be his tartan-clad buttocks in a crude impersonation of Carmen Miranda in a kilt.

The ensuing uproar was deafening, spurring Jock on to further extremes of exhibitionism. Sizing up the situation perfectly, the 'orchestra' drummer quickly handed out maracas and shakers to his three colleagues, forming them into an instant Latin American rhythm section to accompany our blissfully hapless pied piper as he descended from the bandstand and led the hysterical revellers round and round the Heladería in a hilarious, bare-arsed, bagpipe Conga.

Suddenly, it was Hogmanay in Andratx.

MAÑANA

We could hardly have wished for a happier start to the New Year, and as the weeks of January rolled by, we were glad to put the tribulations of that first, trying December behind us. The good times had far outweighed the bad, and although the old house had given us some distinctly unpleasant surprises, we felt that we had coped well enough with the problems that had been thrown our way. We had passed the initiation tests, and we now believed that, at last, we could truly call Ca's Mayoral our home.

Even the long-awaited delivery of our household trappings from the UK was eventually made, though not exactly in the way we had expected. The pantechnicon travelled the final leg of its epic journey across two seas and half the length of Europe only to find that it was too long to negotiate the little bridge in the village, so we were obliged to muster every form of mini transportation, from our own tiny tractor and trailer to Pep's mule and cart and even a fleet of neighbourhood wheelbarrows and hand carts, to lug our stuff, item by item, all the way down that longest last mile home. There wasn't much that the entire surrounding population didn't know about our personal belongings at the end of that day.

Despite our earlier misgivings, Charlie took to his new school like a duck to water, the appeal of wearing a 'uniform' of T-shirt, jeans and sneakers transcending any social or academic teething troubles that we might have anticipated, and he was soon talking in that mid-Atlantic drawl which somehow becomes common to kids of all tongues when they are herded together in international schools. Even his undying love of football was rapidly replaced by a passion for, of all things, basketball – a game he had previously thought fit 'only for girls'. But it suited me just fine; hanging a basketball hoop at the back of the house was, after all, a whole lot easier than trying to improvise a football pitch in an orange grove. And as Charlie got to know more of the village kids, the fun that boys can discover in the great outdoors in such a climate and in such an environment inevitably meant that his beloved TV was soon abandoned, too. 'Can't get anything decent on the poxy thing, anyway,' was his final word on the subject.

Sandy, thanks to the connections made by the Andratx Maradona, did join up with the little football team at La Real, their evening training sessions and Sunday matches quickly providing him with a new circle of friends and a good grasp of the Spanish language – though not all of the vocabulary would have been considered acceptable in the best of mixed company. His initial flirtation with the little *Barbieri* tractor never did develop into a true romance, however. In fact, as the temperatures rose with the lengthening of the days, Sandy's frustrations at being hauled along by the miniature iron horse while ducking and swerving under the clawing branches increased apace. On more than one occasion, I noticed him staring down at the little chugging machine, then looking up longingly at a jet plane climbing away northwards from Palma airport – to Scotland, where the fields had no fruit trees and were large enough to allow *real* tractors to work, perhaps? I wondered . . .

Ellie and I were now well used to our new daily routine of orange and lemon picking, lugging our one-legged triangular ladders from tree to tree in the pleasant winter sunshine, and filling crate upon crate until we had completed our daily orders for Jeronimo the fruit merchant. We were learning the business *poco a poco*, but although Jeronimo always generously paid us the top market prices of the day, it didn't take us long to realise that fruit farming on this scale was never going to be a lucrative occupation – not when we were earning little more for a kilo than the price of one orange in Britain. But then, we were finding that we didn't seem to need much money to get by and live very well in Mallorca, anyway.

Jordi, too, became a valuable source of money-saving tips and information for us. He really was an expert on just about everything we needed to know locally. It transpired that he had a small *finca* of his own – a couple of narrow *bancales* of land up on the slopes of the Coll d'Andritxol – and as he was a man of modest means with little to his name except his old Mobylette, a couple of goats and a dilapidated little tractor, farming cost-effectively was not merely important to Jordi, it was a matter of life and death. He willingly passed on the benefits of his experience to us without any thought of reward, being happy just to meet me outside the Bar Nuevo in Andratx on market days for a seat in the sun, a beer or two and a good go at his favourite hobby – swearing in English.

And what of the Ferrers' cats and dogs? We stood by our promise, and every weekday took food to fill their cluster of bowls outside the Ferrers' *casita*. But rather than ever give in to boiling up Francisca's nauseating poultry off-cuts (and in the absence of a philanthropic butcher who might be prepared to give us free scraps), we actually resorted to buying special dogs' sausages to supplement Francisca's broken rice and any left-overs which we could save from our own table. It was a diet that the animals appeared to thrive on, but to this day I wonder if Francisca's early donation of hens' heads and feet was no more than a ploy contrived to sicken us into supplying victuals for *her* pets at *our* expense. Whether or not, we could never let the poor animals suffer for the suspected perfidy of their mistress.

Our concern for their welfare was eventually rewarded by Robin and Marian who, in their own good time, decided to honour us occasionally with their company, and would even sit on our knees during a break from fruit picking in the orchards – especially if they thought a titbit or two might be on offer. But those social visits ceased abruptly the moment Tomàs and Francisca arrived on Friday evenings to the dogs' howling, yelping welcome. For the rest of the weekend, Robin and Marian would not even glance in our direction, even if we passed within a few inches of them.

As for the cats, they simply kept themselves to themselves and ignored us completely, save for an ill-tempered chorus of gurning and spitting in our direction when we were feeding them. Those cats really did resent our presence, and unlike the dogs, who deigned to allow us permission to pat their heads and tickle their ears from Monday to Friday (insurance, maybe, against the risk of their true masters failing to return), the motley moggies would brook no such condescension. They hated the sight of us, and no amount of 'puss-puss, here-kitty-kitty' peace-making overtures from us was ever going to change that. So be it, then. Our day would come.

Our relationship with the Ferrers themselves settled into one of polite waves in passing, brief exchanges of small talk about tractors and trees with Tomàs over the march wall, or a few 'Oohs' and 'Aahs' about the weather from Francisca when she came over with her bags of broken rice – or, rather, to get her prying nose back inside the front door, as Ellie put it.

Las Calmas de Enero, a spell of calm, summer-like weather with which the island is invariably blessed in January, eventually announced the passing of winter (something of a misnomer for that often most benign and agreeable of Mallorcan seasons) and the coming of spring. Within a few days, the countryside appeared covered in drifts of almond blossom, the delicate flowers of white and palest pink adorning the trees even before the appearance of the leaves, and reminding us of our very first day at Ca's Mayoral, when that rare fall of snow clung to the twigs and branches of those same gnarled trees and blanketed the valley in a magical mantle of white.

It already seemed so long ago, and we were beginning to realise that the passage of time through the unhurried serenity of Mallorcan country life was something to be marked by the slowly-changing face of nature, not by the hands of a clock or the pages of a calendar.

I mentioned this to Ellie late one balmy spring evening when we were sitting on the two rickety chairs by the old well on the terrace, sipping wine and gazing at the stars through a leafy canopy of twisting grapevines. A sleepy chorus of chirping crickets and the lazy croaking of an amorous frog drifted gently up through the trees from the *torrente*, lulling the citrus-scented air with the soothing sounds of the Mediterranean night.

'You know what I think?' she murmured, 'I think you've finally learned how to be *tranquilo*. And I like it. It's, well . . . nice.'

I leaned back in the old chair, savouring the warm tranquillity of the slumbering valley, and marvelling at the silent majesty of the dark mountains silhouetted against the deep velvet-blue of the sky.

One day soon, old Jaume's friend, Pepe the tree *maestro*, would be around to work his much-needed magic on our orchards, and although we knew that we still had many new skills to learn and a lot of hard work to do before the little farm would ever be restored to anything like its former glory, we were ready for the challenge. Jaume had said that everything would start to look better in the spring, and now I was beginning to agree with his opinion a little more with each passing magic moment. Why, I might even buy a pig tomorrow, if for no other reason than to keep old Maria Bauzá happy. Maybe a few hens too . . . I'd think about it . . . *mañana*.

From his secret place somewhere deep in the starlit orange groves, a nightingale began to pour out his silver song in trilling cascades and blithe crescendos, adding the final priceless brush strokes to this entrancing scene.

Good Vibrations
Coast to Coast by Harley

Tom Cunliffe

'. . . A motorcycle the size of "Betty" would have been beyond the dreams of the craziest pack leader at the Ace Café on London's North Circular Road in the monochrome days of Rockers, Nortons and Marianne Faithfull, when good was middle class, bad was misunderstood and the motorcycle offered the stark truth to a generation of inarticulate searchers . . .'

Two Brits, Tom Cunliffe and his wife Ros, take life in the saddle and on the road, and embark on the journey of a lifetime across America astride the quintessential 'dream machine' – the Harley Davidson.

From trailer park culture to Psycho-style motels and an encounter with Hurricane Bertha, they meet an eclectic collection of people; moonshiners, ol' blues men, swaggering cowboys and cowgirls, Sioux Indians, travelling fiddle salesmen, Korean War veterans and the 'hickest hillbilly in the west' (a cockney from Aldershot), each representing a portion of the diversity of modern America.

A fascinating view of America through British eyes and between the handlebars of the great Harley Davidson.

Paperback

The Gringo Trail

Mark Mann

'. . . there I was in the middle of Bogotá, coked up to my eyeballs, in a hallway holding two machetes, while some drunk Colombians argued about whether or not to blow up a bar with a live hand-grenade . . .'

Asia has the hippie trail.
South America has the gringo trail.

Mark Mann and his girlfriend Melissa set off to explore the ancient monuments, mountains and rainforests of South America. But for their friend Mark, South America meant only one thing . . . drugs.

Sad, funny, shocking. The Gringo Trail is an On The Road for the Lonely Planet generation. A darkly comic road-trip and a revealing journey through South America's turbulent history.

Drama and discovery. Culture and cocaine.
Fact is stranger than fiction.

Paperback

Running A Hotel On The Roof Of The World
Five Years In Tibet

Alec Le Sueur

The Holiday Inn, Lhasa, would have proved any hotel inspector's worst nightmare. An hilarious behind the scenes look at the running of an unheated, rat infested and highly confused hotel set against the breathtaking beauty of the Himalayas.

• Highly entertaining but also illuminating and informative

• No other foreigner has spent so long in Tibet since the days of Heinrich Harrer

• Le Sueur provides a fascinating insight into an intriguing country which so few foreigners have been permitted to visit.

Paperback

For a current publishing catalogue and full listing of Summersdale travel books, visit our website:

www.summersdale.com

Suffer from wanderlust? Interested in adventure? Planning to take off?

Go travel or read travel. We've got everything packed into

www.travel-bookshop.com